Heritage Building Information Modelling for Implementing UNESCO Procedures

T0187889

The main aim of this book is to develop and explore the value of new innovative digital content to help satisfy UNESCO's World Heritage nomination file requirements.

Through a detailed exploration of two BIM case studies from Jeddah, Saudi Arabia, the book uniquely connects the use of Heritage BIM to the documentation methods used by UNESCO and demonstrates how this provides a contribution to both countries with heritage sites and UNESCO as an organisation. The research and practical examples in the book seek to address both the lack of a comprehensive method of submitting a nomination file to UNESCO and the lack of authentic engineering information in countries where extensive heritage sites exist. It looks at answering the following questions:

- How can Heritage Building Information Modelling (HBIM) be used to better maintain, protect, and record the updated information of historical buildings?
- How can HBIM provide innovation in creating the missing information for the assignment of UNESCO's World Heritage status?
- What additional value can a sustainable update of HBIM data provide for such sites?
- How can HBIM improve the cultural value of heritage buildings in the short, medium, and long term, as well as provide a better future for historical buildings?

This book will be useful reading for researchers and practitioners in the areas of heritage conservation, archaeology, World Heritage nomination, HBIM, digital technology and engineering, remote sensing, laser scanning, and architectural technology.

Ahmad Baik is Architect and Assistant Professor at King Abdulaziz University, Faculty of Architecture and Planning, Saudi Arabia, Jeddah. He holds a Ph.D. in BIM from University College London (UCL), a Master's of Geospatial Information from RMIT University, Australia, and a Bachelor of Architecture from King Abdulaziz University. He is a BIM specialist with progressive experience in architectural design, 3-D laser scanning, modelling and simulation, virtual reality, and augmented reality.

Heritage Building Information Modelling for Implementing UNESCO Procedures

Challenges, Potentialities, and Issues

Ahmad Baik

Routledge
Taylor & Francis Group

LONDON AND NEW YORK

First published 2021
by Routledge
2 Park Square, Milton Park, Abingdon, Oxon OX14 4RN

and by Routledge
52 Vanderbilt Avenue, New York, NY 10017

Routledge is an imprint of the Taylor & Francis Group, an informa business

© 2021 Ahmad Baik

British Library Cataloguing-in-Publication Data
A catalogue record for this book is available from the British Library

Library of Congress Cataloging-in-Publication Data
Names: Baik, Ahmad, author.
Title: Heritage building information modelling for implementing UNESCO procedures : challenges, potentialities, and issues / Ahmad Baik.
Description: Abingdon, Oxon ; New York, NY : Routledge, 2020. | Includes bibliographical references and index.
Identifiers: LCCN 2020006638 (print) | LCCN 2020006639 (ebook) | ISBN 9780367477981 (hardback) | ISBN 9781003036548 (ebook)
Subjects: LCSH: Historic preservation—Data processing. | Historic sites—Conservation and restoration—Data processing. | Building information modeling.
Classification: LCC CC135 .B345 2020 (print) | LCC CC135 (ebook) | DDC 363.6/9—dc23
LC record available at https://lccn.loc.gov/2020006638
LC ebook record available at https://lccn.loc.gov/2020006639

ISBN: 978-0-367-47798-1 (hbk)
ISBN: 978-0-367-51071-8 (pbk)
ISBN: 978-1-003-03654-8 (ebk)

Typeset in Goudy
by Apex CoVantage, LLC

Contents

Figures

1 Introduction

1.1 Introduction

1.1.1 UNESCO World Heritage Sites and nomination files

The United Nations' Educational, Scientific and Cultural Organization (UNESCO) is a specialised agency, which was founded more than 50 years ago under the United Nations system. The UNESCO constitution, which was approved at the London Conference in November 1945, states that: "Since wars begin in the minds of men; it is in the minds of men that the defences of peace must be constructed". According to Marshall (2011), UNESCO is seeking to "encourage the identification, protection, and preservation of cultural and natural heritage around the world considered to be of outstanding value to humanity". This statement is personified in a global agreement, described as "the convention concerning the Protection of the World Cultural and Natural Heritage". Therefore, the first step to achieving the UNESCO vision is to be in the UNESCO World Heritage List.

Prior to being incorporated into the World Heritage List of UNESCO, the sites must be considered to be of outstanding universal value (OUV) and meet at least one of the ten selection criteria, six of which are cultural and four are of natural criteria. Any country who wishes to add its site needs to go through UNESCO's process. This process encompasses several steps, which is referred to as the nomination process. Chapter 2 (sections 2.2, 2.2.1, and 2.2.2) will provide more details regarding the UNESCO World Heritage and the nomination file (WHNF).

As with many systems in the world that have achieved significant recognition, the UNESCO World Heritage nomination system is facing several challenges and issues. For example, in accommodating the huge numbers of heritage sites around the world, and the lengthy time required to prepare both the site and application file, and sharing increasingly complex data and communicating the information contained in the data between those involved in heritage projects (Bolla et al., 2005; Rao, 2010; Szabó, 2005).

These issues and challenges will be discussed in more detail in chapter 3, as well as presenting a number of solutions that have been developed.

1.1.2   *Historic Jeddah and the nominations file*

Jeddah is one of the most important cities in the Kingdom of Saudi Arabia. It is located in the western region of Saudi Arabia with a coastline on the Red Sea and is considered the economic and tourism capital of the kingdom (Figure 1.1). The city has a long history as it dates back to 3,000 years ago when groups of fishermen settled in the region after their fishing trips around the sea (Al-Fakahani, 2005).

Many of the buildings in Jeddah were built over 300 years ago, such as the "Nasif House" and "Jamjoum House" (see Figure 1.2). These buildings have special characteristics that originate from different aspects of Islamic culture (Telmesani et al., 2009).

Over the last decade, the government of Saudi Arabia has paid huge attention to the city and the Old City of Jeddah. The first goal they set to achieve was for the old city to be included in the UNESCO World Heritage List.

Historic Jeddah's World Heritage nomination file has been submitted more than once during the last ten years; however, the nomination has not been

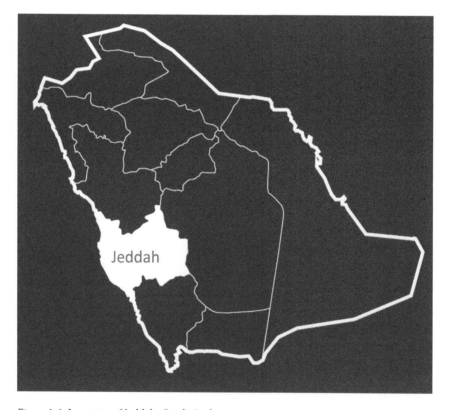

Figure 1.1 Location of Jeddah, Saudi Arabia
Source: Author

Figure 1.2 Nasif House (left side) and Jamjoum House (right side) in Historic Jeddah
Source: Author

successful. Interestingly, the files were not rejected but rather withdrawn by the government of Saudi Arabia. The main reasons for withdrawing the last nomination file of Historic Jeddah, as Ziyad Al-Dirais, UNESCO's kingdom representative explained, was that, "Our dismissal of the Kingdom's application was purely down to technical reasons, as Jeddah's historical sites had been subjected to negligence and misuse from people who did not recognise its value. Moreover, the negligence had greatly influenced ICOMOS' decision". In 2014, at the 38th session of the World Heritage Committee in Qatar, the nomination file of Historic Jeddah, the Gate to Makkah, was eventually approved by UNESCO but many conditions and requirements had to be met

Chapter 2 (section 2.3) will provide more detail regarding Historic Jeddah's nomination file.

1.1.2.1 *Climatic conditions*

The climate of Jeddah city is always uncomfortable as it has a monsoon-type climate; this is due to the location of the city, which is near to the coast of the Red Sea. The city of Jeddah has erratic and scarce rainfall, which regularly falls mainly during October and April. During rainy seasons, Jeddah always experiences flooding in low-level parts of the city because of poor water management infrastructure and the heavy rainfall during these seasons. Between 50 to 100 mm is the average measure of rainfall per year. The prevailing wind is from the northwest; yet, on the other hand, sandstorms blow from the south for about 30 days a year. The average relative humidity is 80% to 85%, coupled with an average daily temperature of 30°, which creates a very oppressive atmosphere within the town (Climatemps.com, 2015).

1.1.2.2   The population of Historic Jeddah

Until the middle of the 19th century, the original families of Jeddah and the trading families from Yemeni, Indian and East Asia inhabited the old city houses. However, Jeddah's incredible growth due to oil revenue and increasing numbers of pilgrims led to the local residents leaving their traditional abodes and moving to newly built suburbs. Today, a high proportion of single, male foreign workers mainly occupy the historical houses of Jeddah by renting a room (or part of a room) from Saudi landlords.

In 2002, a survey was undertaken within the framework of the King Abdul-Aziz Project, which showed that the total population of the old city was lower. The survey showed that the total number of legal residents of the Old Jeddah city was around 13,000 and that most of them were under 45 years of age. In 2011, the government conducted a new survey, which showed that there were some 35,000/40,000 occupants in the Old Jeddah city and probably some 7,000/8,000 people living within the nominated property limits.

1.1.3   Issues surrounding Historic Jeddah

The major problem that confronts the historical district of Jeddah city today, as pointed out by the director of the historic area, Sami Nawwar, is how the local authorities can preserve such buildings. These buildings face a constant risk of collapse and erosion due to ageing and other human factors, as well as disasters such as fires and flood. Indeed, in the last 20 years, huge numbers of historical houses and buildings in the historical district of Jeddah city have been destroyed, and this has led to the erasing of hundreds of years of culture and history. For instance, in March 2010, the historic district in Jeddah city was hit by the most horrible fire in years, and more than six buildings were burnt down (Alawi, 2013).

According to engineer Sami Nawwar (personal communication, 24 February 2013), "these houses were important historical landmarks", and sadly, "the number of historic houses has declined from 557 to 350" in recent years. Most of these houses were "Class A" historical buildings. Sadly, many owners believe that these buildings are not useful as they cost a lot to maintain, have no material return, and are generally not supported by the government. Therefore, they prefer to abandon them until they completely deteriorate and collapse, so that they can build new towers or commercial structures that generate a better return. The owners of these buildings see this option as more attractive and rewarding than preservation due to the high land prices that can be invested in for much better returns. Unfortunately, this is true; thus, such buildings are not treated properly, and neither the owners nor the government appreciate the true value of these historical landmarks. However, these buildings have the prospect to be maintained and used for many purposes, such as museums for antiques and valuable possessions.

Many heritage experts suggest that these historical monuments must be conserved, preserved, maintained, and reused in a better and more organised way, or else they will lose their historical essence and culture over time. Additionally, the main action necessary for heritage preservation is the documentation of these heritage sites. However, there is presently no official Hijazi architectural database or adequate records in existence due to the absence of specialists and experts to carry out this job. Hence, Alitany et al. (2013a) pointed out that "there is an important need for trained professional and infrastructure to preserve the city and its buildings".

Since ten years ago, according to Nawwar (2013), the municipality of Historic Jeddah city has decided to attempt to preserve and develop the area with the help of independent engineering survey offices. This process often takes a long time and can sometimes provide unreliable information since these engineering survey offices used traditional surveying techniques in order to perform the task. This process often takes up to ten months for completion and incurs high costs.

In 2011, in an innovative attempt from Jeddah municipality to come up with an approach to solve the problem in hand, the city decided to develop certain paths for tourists. Their vision was that by working on improving chosen areas, they would plant the seed for development, which would therefore have a positive effect on the neighbouring areas.

However, this method lacked several key aspects such as remote management, shared databases, and integration with any future building condition updates.

The municipality of Jeddah has contributed to the preparation of a restoration manual for rehabilitation in historical Jeddah. This focused on the architectural and structural elements' characteristics, pathologies, and restoration techniques, as well as proposing solutions.

In 2012, after Historic Jeddah was added to the UNESCO World Heritage site's list, the number of historical buildings increased to 1,447 buildings. A huge number of these buildings were in danger or in need of some conservation, according to Mohammad Yosof Al-Aidaroos, the Supervisor of Archaeological and Historical Sites at the Supreme Commission for Tourism, Saudi Arabia.

1.1.4   *Heritage BIM as a new method for UNESCO's WHNF*

After the United Kingdom government made it mandatory for all public sector projects to adopt Building Information Modelling by 2016, key institutions in the construction industry, such as the Royal Institution of Chartered Surveyors (RICS), have published guidance documents regarding the definition and application of Building Information Modelling (BIM) in the design process. Changes have been made in the associated legal outlines for projects adopting BIM and in the impact of BIM adoption for the building sector.

The story of BIM actually started in the 1970s, under the name of building product modelling (Eastman, 1975). Since 1995, when the International Alliance for Interoperability (IAI) started to develop a standard format to exchange between different software tenders used in engineering sectors, the issue and the necessity for BIM has become a wider industry issue (Kiviniemi, 2011). The development of BIM in the first ten years was in relation to the technology required in order to deliver the concept of exchanging the data within the BIM workflow among different participants (Kiviniemi, 2008). The first and basic concept of applying BIM is to design and build new buildings with high performance in each sector of the building industry, for instance in terms of energy, economic and structural analysis, as well as the scheduling of the works as an addition to these new designs (Eastman et al., 2011).

Recently, employing BIM into the heritage field has been introduced as a new method known as Heritage BIM (HBIM), which can be considered as the fundamental stage toward the BIM workflow for retrofit and reconstruction tasks (Ebim, 2015). This method has been used in small numbers of projects worldwide, such as in the research of Fai et al. (2011a), Murphy (2012), Oreni (2013), and Penttilä et al. (2007). The main objective of these research projects was to provide intelligent data (S. Fai et al., 2011a), and "an as-built" digital 3-D model (Backes et al., 2014), in order for them to be used for several purposes, such as documentation, conservation, and management.

In an attempt to provide a new model of the UNESCO WHNF, Historic Jeddah, Saudi Arabia's file was chosen as a case study for this book. This case study will be referred to as Jeddah Heritage Building Information Modelling (JHBIM). The main reason for choosing this case study is due to the huge gap in the knowledge in relation to heritage buildings in Historic Jeddah. Many of these buildings have no engineering data in order to enable them to be restored and rebuilt in the case of collapse or any disasters. In addition, these buildings need to be examined in the next few years for UNESCO's WHNF annual inspections.

This book will illustrate an interactive approach to moving from the existing out-dated two-dimensional (2-D) systems and three-dimensional (3-D) models to the JHBIM for the purposes of meeting the requirements of UNESCO's WHNF.

This approach will be based on the integration of different data sources within a common interactive environment, as well as the ability to extend hosting all of the data for Jeddah's heritage buildings. The interactive environment can allow users from UNESCO or other involved organisations to extract and generate the information that they require directly from one database. It also allows for the ability to add reports, comments, and enquiries for the participants on the nomination file.

The possible outcomes of employing this method will benefit UNESCO, the involved organisations, and the participants of the nomination file. A few of the key advantages for the participants are the clear time frames and processes to be followed, the reduction in cost and time required to prepare the nomination file, the sharing of information, and the better control and management of the tasks and manpower involved. This will also lead to fewer errors and clashes in relation to the tasks undertaken during the preparation of the nomination file. Additionally, employing this method will provide easy access to the information and make it easier to be modified in case of any issues arising during the preparation of the file, alongside allowing for any feedback from UNESCO's organisations.

The key advantages for the involved organisations (e.g. ICOMOS and IUCN) can include early follow-up with regards to the preparation steps of the file via the Heritage BIM environment, as well as saving time, money, and effort in sending frequent missions to the heritage site. The key benefits for UNESCO are a reduction in cost and the time involved with preparing such files, as well as cutting down on the nomination process time and increasing the credibility of their decisions.

1.2   Aims and targets

The overall aim of this book is to develop and explore the value of new innovative digital content to help satisfy UNESCO's World Heritage nomination file requirements. This will be achieved by generating a highly interactive digital model which can produce complete engineering information and drawings. Additionally, the Jeddah Heritage Building Information Modelling (JHBIM) will also be developed, which will be managed, shared, and remotely reviewed (Baik et al., 2013).

The Jeddah Heritage BIM will depend on the Terrestrial Laser Scanning (TLS) and image survey data (ISD). The output information could be used for several applications and will be available for remote management, creating a historical digital documentation database for maintenance and building renovation efforts.

Practically, a number of historical buildings in Historic Jeddah will be chosen as a model for this book, including the Nasif Historical House (main case study).

Figure 1.3 shows the relation of JHBIM and other engineering sectors.

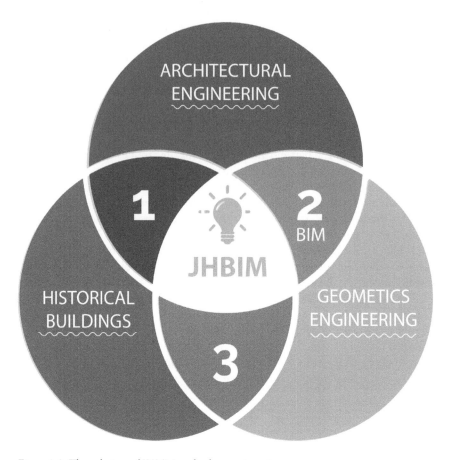

Figure 1.3 The relation of JHBIM and other engineering sectors
Source: Author

1.3   Investigation

Assuming that the HBIM can be used to provide an ideal model for UNESCO's WHNF requirements and to solve some of the issues with preserving the historical buildings in Jeddah city, the following questions arise and will be answered within the context of this book (see Figure 1.4).

The HBIM method has been adopted for a case study of Historic Jeddah, which is described as Jeddah Heritage BIM (JHBIM) and focuses on the Nasif Historical House, which is considered to be one of the most important historical houses in Jeddah.

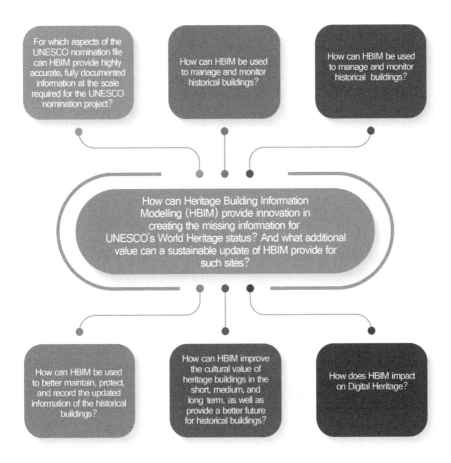

Figure 1.4 Questions can be answered within the context of this book

1.4　Methodology

The methodology will be applied to develop a new way in which to meet the requirements of UNESCO's World Heritage nomination file. This method has been adopted for a case study of Historic Jeddah, which is described as Jeddah Heritage BIM (JHBIM), and integrated and explained over the book's chapters, and outlined in this section.

The terrestrial laser scanning (TLS) and the close-range photogrammetry have been employed as the initial methodology for the purposes of providing data input for the JHBIM model. The framework explains the ideal TLS station locations, the scanning resolution, and the LiDAR point cloud registration processes. Via this step, an architectural survey-framework can be offered. This framework is suitable for different types of heritages to be documented and managed in a more efficient way. Moreover, this method can provide the basis for the inventory plan, which is the most important and fundamental requirement of UNESCO's World Heritage nomination file.

The heritage buildings in Historic Jeddah are similar to many of the heritage buildings all around the world in regards to how they were built and how complicated they are. In fact, many of the architectural elements of these buildings were designed and built on-site in order to fit their exact locations. Because of this, a quest has taken place to provide the Hijazi Architectural Objects Library (HAOL) as well as to reproduce Hijazi elements as 3-D computer models, which are modelled using the Autodesk Revit Family (RFA). The HAOL is dependent on the image survey, point cloud data, and the Hijazi architectural pattern books, such as those provided by Greenlaw (1995). However, this book shows Suakin city to be in Sudan, which is opposite to Jeddah but on the African side. Another general book that covers Arab architecture is *Traditional Domestic Architecture of the Arab Region*, by Friedrich Ragette (2003). This new Hijazi Architectural Objects Library has been developed as a plug-in for the Autodesk Revit platform and inserted into the JHBIM model.

After completing the JHBIM model the next step will be integrating the non-geometric data and any existing information such as restoration reports, GIS data, images and text data into the JHBIM model in order to meet the requirements of UNESCO's WHNF and to provide full engineering drawings and information. This engineering information can be used for any conservation and preservation purposes in the future.

1.5   Overview and summary of the book chapters

Chapter 2: Review of UNESCO's World Heritage Nomination Files (WHNF), Building Information Modelling (BIM), and Heritage Building Information Modelling (HBIM)

This section of the book will introduce and explain UNESCO's World Heritage nomination file (WHNF) requirements by summarising the manual relating to the preparation of World Heritage nominations. Next, the nomination file of Historic Jeddah in Saudi Arabia will be explored as a case study of UNESCO's WHNF, starting with the background of architectural characters of Historic Jeddah. Lastly, there will be an introduction to Building Information Modelling (BIM) and Heritage Building Information Modelling (HBIM).

Chapter 3 : the new model for UNESCO's World Heritage nomination file

This chapter will discuss the issues around preparing UNESCO's WHNF, and why it is important to have new methods for preparing the nomination file, as well as how Heritage BIM can solve the issues by providing all the required information regarding the sites that will be listed in UNESCO's World Heritage Sites List.

Chapter 4: Jeddah Heritage BIM and the case study

This chapter will introduce the case study and the issues that relate to the historic district. Afterwards, employing BIM for the historic buildings in Jeddah, Saudi Arabia, will be discussed in order to solve the issue and to meet UNESCO's WHNF. Throughout this chapter, the architectural style of the Hijazi buildings will be presented in order to represent them in a more realistic way and in order for them to be used as a basis for the modelling structures. Thereafter, the methodological work will be presented. This will include the data captured and the data processing steps. This will be done by following the JHBIM and the HAOL modelling process. The last step will be explaining how the requirements of UNESCO's WHNF will be met through the JHBIM.

Chapter 5: the conclusion

This chapter outlines the summary of the book and the proposed model of using Heritage BIM as a new model for UNESCO's WHNF. Additionally, there will be a summary of the process of JHBIM. Afterwards, this chapter will highlight the limitations and challenges, as well as the contributions of this book.

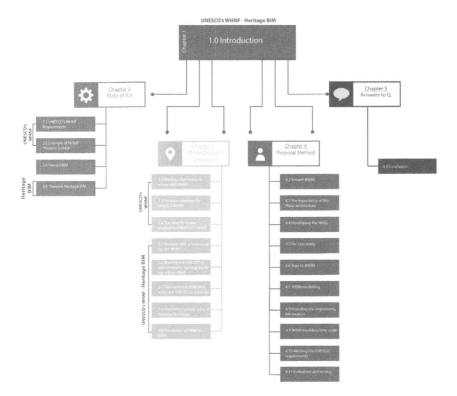

Figure 1.5 Book outline

2 Review of UNESCO's World Heritage Nomination Files (WHNF), Building Information Modelling (BIM) and Heritage Building Information Modelling (HBIM)

2.1 Introduction

Introducing and explaining UNESCO's World Heritage Convention and requirements for the nomination file by summarising the manual of preparing World Heritage nominations will be presented directly following the introduction (i.e. the second part of chapter 2). This includes highlighting the critical importance of Outstanding Universal Value (OUV) that demands to be validated in the nomination files submitted to UNESCO, with reference to inclusion in the list of World Heritage sites. Therefore, all sites or properties submitted for this process are subject to detailed examination of their potential for OUV that relates to their intrinsic value, how sites are planned to be managed, sustained, monitored, conserved, promoted, and protected. The third part of this chapter will explore the nomination file of Historic Jeddah, "Gate to Makkah" in Saudi Arabia as an example of UNESCO's World Heritage nominations file. The aim of the fourth part of this chapter is to introduce Building Information Modelling (BIM), and in the fifth part it is to introduce the development toward Heritage Building Information Modelling (HBIM).

2.1.1 The UNESCO

The UNESCO is the "United Nations Educational, Scientific and Cultural Organisation", which was founded around 70 years ago by the United Nations (UN) and the London Conference in 1945, which reported: "Since wars begin in the minds of men; it is in the minds of men that the defences of peace must be constructed". The major aim of the UNESCO, according to Duncan Marshall (2011, p. 10), the coordinating author of the World Heritage Resource Manual, is "to contribute to peace and security in the world by promoting collaboration among nations through education, science, culture and communication in order to further universal respect for justice and the rule of law and for the human rights and fundamental freedoms which are affirmed for the peoples of the world, without distinction of race, sex, language or religion".

2.1.2 World Heritage background and the mission

According to the official definition, UNESCO World Heritage is seeking "to encourage the identification, protection, and preservation of cultural and natural heritage around the world considered to be of outstanding value to humanity" (UNESCO World Heritage Convention). According to Eléonore de Merode et al. (2004, p. 8), "The fact that this Convention deals with both cultural and natural resources makes it a unique and powerful tool for the protection of heritage". The mission of UNESCO's World Heritage, as the manual of preparing World Heritage nominations states, is as follows:

- To encourage the worldwide countries to sign the World Heritage Convention and to ensure the protection of their natural and cultural heritage; encourage State Parties to the Convention to nominate sites within their national territory for inclusion on the World Heritage List;
- Encourage State' Parties to establish management plans and set up reporting systems regarding the state of conservation of their World Heritage sites; help State Parties safeguard World Heritage properties by providing technical assistance and professional training;
- Provide emergency assistance for World Heritage sites in immediate danger; Support State Parties' public awareness-building activities for World Heritage conservation;
- Encourage participation of the local population in the preservation of their cultural and natural heritage; Encourage international cooperation in the conservation of our world's cultural and natural heritage.

2.1.3 World Heritage nominations

To achieve inclusion into the UNESCO World Heritage Sites List (WHSL), according to the World Heritage Resource Manual, the site needs to represent "an Outstanding Universal Value that should include one of the ten selection criteria" that are "four natural and six cultural criteria".

When a country identifies a site of outstanding universal value (OUV), it needs to begin the nomination process for UNESCO that has several steps, as described in Figure 2.1.

2.2 UNESCO World Heritage nomination file requirements

The most important requirement for UNESCO for the historical sites is to have an outstanding universal value (OUV), and include at least one of the ten criteria. In addition, integrity and authenticity are important factors for these historical sites, together with the requirements for management and protection of these sites. Finally, procedures for monitoring these historical sites must be provided (Santana-Quintero and Van Balen, 2009).

These requirements are detailed in Table 2.1.

UNESCO WHSL Nomination Process

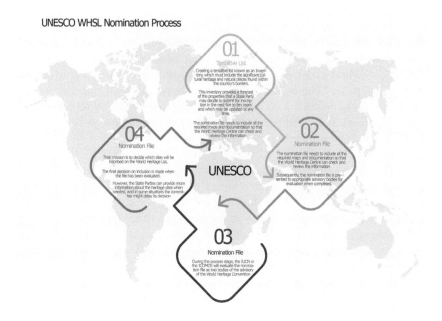

Figure 2.1 UNESCO nomination process

Table 2.1

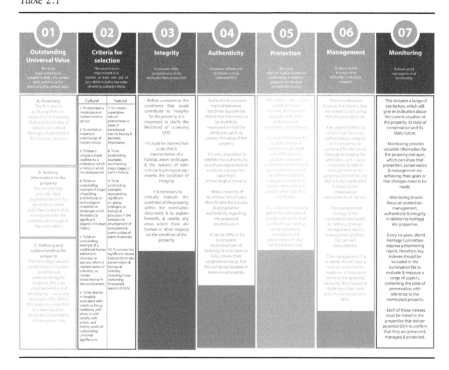

2.2.1 Evaluation process

After the nomination file has been submitted, including all the required data, the procedure of evaluation begins with the following stages as shown in Figure 2.2.

It could be noted that after the advisory bodies have evaluated the nomination file and before the Committee considers the file, the "advisory bodies" have the

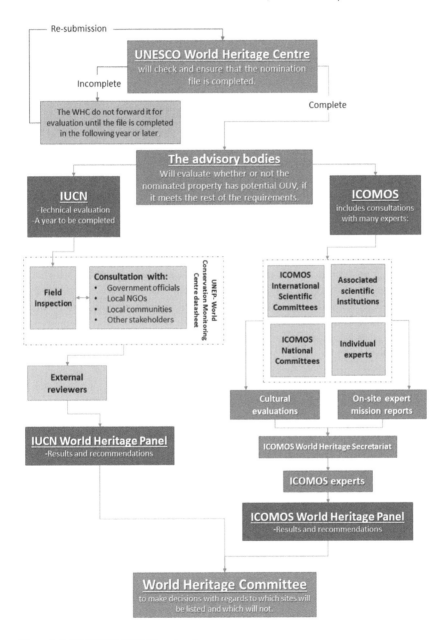

Figure 2.2 UNESCO WHNF evaluation process

right to ask questions or request more information from the State Party by 31 January of the submission year. By 28 February, the requested information must be received by the Advisory Bodies to be considered. Prior to the Committee meeting, the state party is required to communicate with the Chairperson to highlight any mistakes in the evaluation of the advisory bodies. The state party must advise the World Heritage Centre if there are any developments during the evaluation that could affect the nominated property. The final step of this process is the decision by the Committee of the World Heritage with regards to whether the nominated property would be listed or not. These decisions by the committee are normally supported by recommendations made by the relevant advisory bodies.

2.2.2   Inscribing the site in the WHL

After inscribing the heritage site into the World Heritage List (WHL), a number of procedures are put into action, some directly and some indirectly. The first direct action is providing UNESCO and ICOMOS with any requirements or missing information. The second action is completing the protection and management plan. The third action is the annual on-site evaluation and inspection. In this step, UNESCO and ICOMOS send teams of experts to examine and evaluate the site and prepare an annual report regarding the site.

The examination and the evaluation focus on the management plan and the processes of protecting the World Heritage site, which were provided within the nomination file previously (Saudi SCTA, 2015).

In case the site fails any of the tests, the consequences could lead to much accountability, and it could also mean delisting it from UNESCO's World Heritage List after being qualified for the OUV stage. This previously happened to the Arabian Oryx Sanctuary in 2007 when it was delisted from the WHL after the government increased the oil extraction activity in the area and reduced the sanctuary by 90%.

The second case of delisting a site from UNESCO's World Heritage List was done in 2009 to the Dresden Elbe Valley, Germany, which was done as a result of the building of the Waldschlösschen Bridge, as according to UNESCO's World Heritage Committee. The bridge has had an ongoing serious impact on the integrity of the site's landscape.

2.3   The example of UNENSCO's WHNF: Historic Jeddah, Gate to Makkah

This section will explore the background and the architectural and urban characteristics of Historic Jeddah in Saudi Arabia. Indeed, Jeddah city is one of the most important cities in the Kingdom of Saudi Arabia; it has a long history and many historic buildings that were constructed more than 300 years ago.

Two of the most important houses in Historic Jeddah are Nasif Historical House and Noorwali Historical House. These houses will be presented in this chapter to give an example of these magnificent buildings, including the architectural characteristics of these houses, such as the Roshan, Mashrabiyah, and the Manjur. Lastly, the chapter will present and summarise the Historic Jeddah nomination file, "Gate to Makkah", which was presented to the UNESCO World Heritage Centre and the related issues surrounding it.

2.3.1 The Old City of Jeddah: background

The city of Jeddah has intrigued explorers for a long time. Many Arab and Muslim travellers, as well as authors, have noted that the city's name derives from the Arabic name for "grandmother", and this is because the area known as the cemetery of Mother Eve is located in the northeast of Jeddah. However, other authors have pointed out that the name is related to "Quda'a," an Arab tribal chief, who migrated from the south to find a new homeland 2,500 years ago (Al-Fakahani, 2005).

The city of Jeddah continued to rise due to fishing and trade, and over time, since the Islamic era's rise, the city has become more important as the main gateway in order to reach the holy city of Makkah, particularly after the visit of Caliph Othman Bin Affan in the year 26 Hijrah (624 AD) (Telmesani et al., 2009).

Jeddah maintained its position as a gateway to Makkah and Madinah for Hajj and due to the possibility of visiting the Prophet Mohammed's grave and his mosque in Madinah (Pesce, 1974).

The association with the Hajj season every year has been the primary reason for the increase in commercial and trading activities, and as a result, has aided a development in the social, political, and profit-making potential of Jeddah city. During the Hajj season, many pilgrims from different countries around the world have visited the city; others have stayed, and Jeddah has grown in size as a result. Furthermore, the reputation of Jeddah city continued to grow. From 1177 AD, the city was faced with significant political and economic insecurity when the city turned out to be part of Ayyubid County, mainly due to the internal conflicts among Ayyubid's princes.

Later in 1254 AD, Jeddah came under the ownership of Mamluk sultans from Egypt for the next two centuries. During the Portuguese Aggression period, the Sultan, Qansuh Al-Ghuri, took defensive measures against the Portuguese who were attempting to seize control of the trade routes in the Red Sea; these included building cisterns, constructing fortified walls, and digging water wells for the city of Jeddah in order to protect the city from any possible siege.

However, all these measures were unsuccessful in stopping the Portuguese navy from attacking ships that were carrying trade goods and pilgrims to Jeddah, and its harbour faced a siege at several points. During 1517 AD, the power of the Mamluk sultans had declined and the Ottomans took over the region of Hijaz (Angawi, 1988).

In 1525 AD, the Ottomans fortified the walls of the city and began to reconstruct its buildings after defeating the Portuguese army. For the next four centuries, the Ottoman Empire controlled the Hijaz region, which included Makkah, Madinah, and Jeddah. However, the local authority was delegated to Al-Ashraf, descendants of the Prophet Mohammed (peace be upon him) (Bokhari, 2006).

According to Al-Fakahani (2005, p. 20), "Although the economic stability of Jeddah varied throughout these years, particularly after the Suez Canal was opened in 1869, it was able to consolidate its trading position in the area". Many countries opened diplomatic missions in Jeddah such as Britain, Holland, and France, and then later, the USA.

In 1915, Al-Shareef Hussein Bin' Ali revolted against the Ottoman Empire by attacking the controlling authority of Al-Ashraf of Hijaz to create the Arab Kingdom. However, nine years later, Al-Shareef Hussein Bin' Ali surrendered, and he turned over the Hijaz region, including Jeddah, to King Abdul-Aziz ibn Saud, who formed the modern country of Saudi Arabia. Since this period, the city's old walls have been demolished and the harbour expanded, which allowed Jeddah to develop and grow to over 1,000 km², and hold a population of 2,500,000 residents.

2.3.2   The architectural characteristics of Historic Jeddah buildings

Little is known about this topic, and a survey of this historic district could be useful in presenting how to effectively restore this district. In fact, the historical houses of Jeddah are rich in architectural characteristics, such as the Roshan, Mashrabiyah, Manjur Pattern and Plaster decoration (Figure 2.3). These architectural characteristics are known as Hijazi architecture style (*Hijazi* is an Arabic word, describing the West region of Saudi Arabia). According to Ragette (2003, p. 30), "these historical houses have remarkable and simple design and architecture

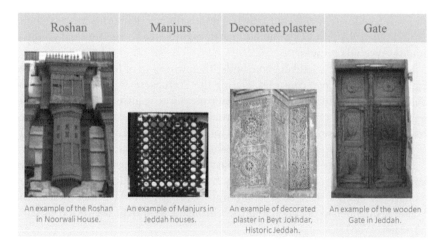

Roshan	Manjurs	Decorated plaster	Gate
An example of the Roshan in Noorwali House.	An example of Manjurs in Jeddah houses.	An example of decorated plaster in Beyt Jokhdar, Historic Jeddah.	An example of the wooden Gate in Jeddah.

Figure 2.3 The common architectural characteristics in Historic Jeddah buildings

that represent a rich heritage, demonstrating how local craftsmen and builders adapted designs to respond to social demands and other environmental factors in earlier periods". As a consequence of this evolution, the designs of the old houses have a unique pattern, as well as being authentic and functional. Moreover, the uniqueness of pattern takes a turn for the better in reducing humidity and increasing thermal comfort, as the buildings increase cross-ventilation (Eleish, 2009).

Moreover, according to SCTA (2013, p. 49), the historical houses of Jeddah have to be "understood as an urban unit active in the making of the city", therefore these houses need to be "studied as typo-morphological responses to climate, material and socio-spatial practices". The basic and primary urban unit of the historic Jeddah houses is the Roshan. These Roshans had a significant role in the shaping of the urban fabric, which originally comprised tightly knit areas integrating commercial and residential functions, organised around the main market and the social identity of the historical city. Furthermore, SCTA (2013, p. 49), pointed out, "Through its programmatic, climatic, spatial and visual characteristics, it contributed to the shaping of the urban morphology, land use patterns and the overall character of Jeddah".

In recent years, the historical houses have been used as multi-purpose buildings that house residential and commercial activities. Additionally, at street level, the historical houses are domestic private spaces combined with commercial semi-private spaces such as offices, warehouses, and sometimes as hotels during the Hajj (SCTA, 2013). Regarding the climatic considerations, historical Jeddah houses are very effective, which was critical in shaping the morphology and the urban fabric of the historic city. Furthermore, the street network of the historic city corresponds to the prevailing breezes, north and northwest. Lastly, shade and light, the alternation of cool and warm surfaces, and hot and cool spaces have an effect on the airflow of the city.

The historical houses, in general, could be described as independent or semi-independent units. In such a humid and hot area, this produces more comfortable streets and increases air flow and cross-ventilation. The proximity of the historical Jeddah houses and their height are also aspects that keep the streets protected from heat and sunrays, and keep them in the shade. In addition, the high houses are used as "wind catchers" in order to allow the sea breeze to maintain continuous vertical air circulation inside the houses. The natural upward movement of hot air across stairwells and shafts pulls air over the windows (Roshan and Mashrabiyah), which in turn cools the inside and favours air circulation.

An important architectural solution for dealing with Jeddah's climate is "Al-Mabit" in Arabic, which was used for sleeping during the summer nights; it was normally built from panelled wood with louvers and a light roof. Moreover, Al-Lyaly (1990) defines it as, "Al-Mabit on the uppermost floor is like an air pavilion". "The louvered timber walls surrounding it on two or sometimes three sides allow the air to circulate freely in the space and at body level thus enhancing the comfort of the occupants".

However, there is a huge gap in presenting these architectural characteristics in digital format to be used for different purposes. For example, presenting these

Hijazi architectural characteristics in a digital parametric form can be useful for documentation, preservation, conservation projects, and can be used for any future projects in the Historic District.

2.3.3 The main houses of Historic Jeddah

Since the 1970s, old Jeddah houses have attracted experts and researchers from all around the world, mainly due to the fact that the Jeddah Municipality has sponsored a number of projects and studies in relation to the historic city. These studies were facilitated for the purposes of the old city's upgrading and conservation. The work that was done by Sir Robert Matthew was the first and most important one. The work he has done was in terms of setting up the basic styles of the current buildings in the historic district, which included making a list of all the historic buildings in the district. These buildings were classified into three levels based on the building's historic significance and the architectural characteristics (see Figure 2.4). Class A was used to describe buildings of national significance, Class B was for regional, and Class C of local significance (Matthew and Johnson-Marshall, 1980).

Today, the Jeddah municipality uses the Geographic Information System (GIS) to survey and record the historic district's buildings and plots. Moreover, the main step in the GIS was updating the classification map for the Class A, B, C buildings. This map was originally designed by Sir Robert Matthew by inserting all the information in relation to the building's conditions. Furthermore, according to the municipalit;, they aim to use the GIS as a key tool for city management and for monitoring all the changes in the district (SCTA, 2013).

Figure 2.4 The historical buildings' classification map by Matthew and Johnson-Marshall (1980)

2.3.3.1   *Nasif Historical House (a Class A building)*

Nasif House, one of the most famous houses in the historic area, was built at the end of the 18th century by Sheikh Umar Effendi al-Nasif, the agent of Jeddah for the Sharif of Makkah (Figure 2.5). The main reason for the house's reputation is due to the fact that during the time when King Abdul-Aziz entered Jeddah in 1925, he stayed in Nasif House, where he met the most notable people of Jeddah as the house was supposed to be the most distinguished and appropriate residence for the Sultan. The architectural style of the house is similar to other historic Jeddah buildings with Roshans and Mashrabiyahs.

Until 1975 the house belonged to the Nasif family. One of the heirs, Sheikh Muhammad, turned Nasif House into a private library that contained more than

Figure 2.5 Nasif Historical House
Source: Author

16,000 books. Today, Bayt Nasif has been restored and has become a museum and cultural centre (SCTA, 2013). Section 4.3 will explain more about the house.

2.3.3.2   *Noorwaloi House (a Class A building)*

The house is located at the beginning of Al Alawi Street; this street was one of the most famous streets in historic Jeddah. It became famous due to the presence of the markets and shops on this street. Noorwali House was built 150 years ago and the house contains 45 rooms. The Noorwali family owns the house. This family is known as a rich family with Indian origins. The family's culture origin had an impact on the style and decoration on the façade of this house. According to Sheikh Abdul-Hameed Noorwali, today, the value of the Noorwali House is more than 80 million Saudi Arabian Riyals (SAR), which is about 15 million (GBP). This value is just for an area of around 300 m², giving the house a high economic and social value. The unique architectural characteristics of this house are the secret of its beauty; in fact, the house has the most beautiful Roshans (windows) in historic Jeddah. This house is also known as "the Green House", due to the green colour of the Roshans.

2.3.4   *Historic Jeddah nomination file (the Gate to Makkah)*

The "Gate to Makkak", the nomination file that was accepted by UNESCO, was compiled by the Saudi Commission for Tourism and Antiquities (SCTA). This gate is the defining pathway to the holy city of Makkah, which extends over an area of 179,000 m² and contains about one-third of the area originally circled by the old city walls, which were removed in 1947.

Furthermore, the property covers the central area of the old city, as well as the areas of the three historic quarters: Yemen, Sham, and Mazloum. In addition, according to SCTA (2013, p. 2), the nominated property "includes the ensemble of the preserved urban fabric of the old city, East of Dahab Street, till the historic limits of the old city to the East" and it has "an elongated shape with a maximum length of about 1,000 meters and a maximum width of around 600 meters".

Three main axes, which are called "the historical paths", have been developed for the nominated property, which are the historic souks in the West-East, and a North-South commercial spine connecting the Gate of Madinah with the southern side of historic Jeddah. It is entirely encompassed with a wide buffer zone that covers the remaining zones of historic Jeddah and the neighbouring residential and commercial zones (SCTA, 2013).

Figure 2.6 shows the georeferenced image showing the limits of the nominated property and buffer zone.

2.3.4.1   *Providing WHNF requirements within the current Historic Jeddah nomination file*

The government of Saudi Arabia had provided the requirements of the World Heritage nomination file in 2012 with more than 697 pages; these requirements are summarised in Table 2.2, Table 2.3, Table 2.4, and Table 2.5.

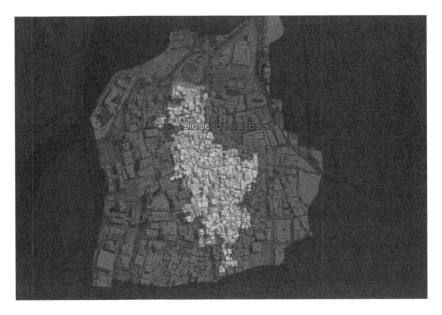

Figure 2.6 Georeferenced image showing the limits of the nominated property and buffer zone—GIS Orth photography satellite image 2007

Table 2.2

	1- Statement of the OUV
Historic Jeddah (The Gate to Makkah) WHNF requirements	The Gate to Makkah, historic Jeddah, is described as an "Urban property", covering an area of 17 hectares in the centre of old Jeddah city.
	A large buffer zone is surrounded by the nominated property, made up of around 113 hectares.
	Old Jeddah city illustrates a marvellous improvement for the architectural tradition of the Red Sea; moreover, on both coasts of the Red Sea, the constructive style is common within the cities along them.
	According to SCTA, "Only a few numbers are preserved outside the Kingdom of Saudi Arabia and the nominated property".
	Due to the Muslim annual pilgrimage, historic Jeddah, which is the gate to the holy city of Makkah, has always been attractive for Muslims from around the world in which to reside and work, as well as this contributing to the city's growth.
	The Outstanding Universal Value of historic Jeddah, the Gate to Makkah, according to SCTA, relates to "Its unique development of the Red Sea architectural style, to its preserved urban fabric, and to its symbolic role as a gate to Makkah for Muslim pilgrims reaching Arabia by boat throughout the centuries".
	The large and complex wooden casements which decorate the historic Jeddah tower houses are an outstanding evolution in relation to the lower coral houses that have characterised the majority of the cities on the two coastlines of the Red Sea since the 16th century.

Table 2.3

Historic Jeddah Nomination File

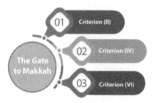

01 Criterion (II)

02 Criterion (IV)

03 Criterion (VI)

The Gate to Makkah

01 According to SCTA (2013), the cityscape, with reference to the Gate to Makkah, historic Jeddah, "is the result of an important exchange of human values, technical know-how, building materials and techniques across the Red Sea region and along the Indian Ocean routes between the 16th and the early 20th centuries". Also, the international sea trade was the main source of representing the cultural world by possessing a shared geographical, religious, and cultural background. It also contains building settlements with innovative and specific technical and aesthetic solutions to cope with the extreme humidity and hot climatic conditions of the region.

Over many centuries, Jeddah City was the richest, largest, and most important among all the Red Sea settlements. Further, it is the last surviving urban site still preserving the ensemble of the features with reference to this culture, multi-cultural environment, commercial-based economy, coral masonry construction, isolated outward-oriented houses, specific technical devices to favouring internal ventilation, and precious woodwork decorating the façades (SCTA, 2013).

03 Historic Jeddah, the Gate to Makkah, is directly associated, both at the symbolic intangible level and at the architectural and urban level with the Hajj (the yearly Muslim pilgrimage to the holy city of Makkah). For many centuries, the city of Jeddah lived in accordance to the pilgrimage. In fact, for centuries, Jeddah was the landing harbour for all the pilgrims that reached Arabia by sea. During the Hajj season, it was very common that pilgrims brought their goods from all around the Islamic world, especially the west of Africa and East Asia, and sold them in the city's Souks.

The association of the old city with the Hajj is also very evident in the urban structure of the nominated property. This includes the traditional souks running East-West from the sea to Makkah Gate, the Ribats and the Wakalas' that used to host the pilgrims.

The association is also evident in the architecture, notably in the façades and internal structure of the houses, and in the very social fabric of the city, where Muslims from all over the world mingled, lived, and worked together. This encouraged intangible and tangible relationships, demonstrates the intimate and long-lasting connection between the pilgrimage and the nominated property. This is a great example of why the city has such rich cultural diversity, caused merely as a result of this unique annual event in the Islamic world.

02 Historic Jeddah is the only surviving urban ensemble of the Red Sea cultural world. The houses of the Jeddah tower houses are an outstanding example of a typology of buildings unique within the Arab and Muslim world. These houses had specific aesthetic which included decorated Roshan façades, the absence of a courtyard, rooms rented for pilgrims, as well as ground floor rooms used for offices and commerce. The functional patterns are reflected in their adaptation to both the humid and hot weather of the Red Sea and to the specificity of Jeddah.

The gate to the holy city of Makkah was significant for the pilgrims arriving by sea, and an important international commercial port. During the second half of the 19th century, the development of the Roshan tower houses illustrated the evolution of the patterns of trade and pilgrimage in the Arabian Peninsula and in Asia. Since 1869, when the Canal of Suez was opened, this caused a huge development of steamship navigation routes connecting India and East Asia with Europe.

The extraordinary relevance of the houses of Jeddah's tower is further enhanced by the fact that they are not only unique within the Red Sea cultural region, but they are also the sole remnants of an architectural typology that was born in Jeddah and that, at the end of the 19th century, spread to the nearby Hijazi cities of Madinah and Makkah, from where it has since completely disappeared under the pressure of modern development (SCTA, 2013).

Table 2.4

Meeting the UNESCO WHNF requirements via JHBIM

3- Integrity	4- Authenticity
Within the JHBIM proposal, the integrity condition can be attributed with more validity and reliability due to the hugely accurate data captured and high LoD modelling. Besides, the JHBIM can focus on providing the boundaries of the proprieties via the integration with GIS, and offering a preservation and management plan.	Within the JHBIM proposal method, the authenticity is able to be presented via highly detailed documentation records, which can be described as as-built records. These records act as reference points for any preservation or conservation purposes. Moreover, it can be the next step for providing a model of the historic city based on the old photographs that were taken in the first half of the 20th century. This step can give a clear picture in relation to how Historic Jeddah and the Gate to Makkah are parts of an authentic traditional urban environment, capable of conveying an image of what this Red Sea commercial and pilgrimage city used to be. What is more, this step can also give an idea about the amount of destruction and the rapidity of the irreversible changes that have affected parts of the historic district of Jeddah over the last 50 years.

Table 2.5

5- Protection	6- Management	7- Monitoring
The Saudi Commission for Tourism and Antiquities and in coordination with the Jeddah Municipality, as well as the participation of the civil society, have drawn up a general strategy for the preservation and revitalisation of the area.		In the case of historic Jeddah, and based on the SCTA vision, the monitoring allows the record of changes at two scales: the larger cityscape in which the site is located and the actual management of the nominated property.

Historic Jeddah (The Gate to Makkah) WHNF requirements

5- Protection	6- Management	7- Monitoring
	The local branches of the Jeddah Municipality and the SCTA are responsible for the daily management of the nominated property, including organising the protection, maintenance, cleaning, and presentation of the historic site. This traditional mechanism, based on the charismatic figure of the Umdahs, permits reaching an ensemble of the population and to involve merchants' and owners' associations in the management of the property. The Jeddah Municipality has approved new urban regulations that set out precise and strict obligations when dealing with historic buildings and empty plots within the nominated property and its buffer zone and as a result, the preservation of the Outstanding Universal Value of the site is guaranteed. The reduction of the rate of decay of the historic houses is a key longterm requirement for the protection and management of the property. A lot of historical buildings have been abandoned and squatted in by poor immigrants, and the controls of the speculative moves that jeopardise the ensemble of the historic city represent the most relevant priorities.	For an urban site such as historic Jeddah, the Gate to Makkah, three distinct and complementary sets of indicators, ranging from conservation, to social statistics and planning data, to verify the overall impact of the revitalisation and conservation strategy proposed for Historic Jeddah, the Gate to Makkah: urban and architectural conservation indicators, social indicators, planning and development indicators, urban and architectural conservation indicators, and the record of environmental data, offers essential information to be crossed with site deterioration. The development of a high-quality 3-D survey of the historic buildings of the old city, with the funds channelled by the Saudi Government from 2013 onwards, will offer a precious graphic tool that allows an easy and precise verification of the rate of the deteriorations taking place in the different buildings of the old city. The Site Management Unit, in the mediumterm, will establish a systematic inventory of the historic buildings and their state of conservation that will be the basis for the definition of priorities for the conservation of the nominated property and its buffer zone. Regarding the rehabilitation strategy of the built environment of the historic Jeddah, it will see the actual impact via a yearly record of the number of working conservation sites in the city. All the records of damages will be reported to SMU in an informal way via the ensemble of the stakeholders, which will be recorded in the historic houses' inventory.

(Continued)

Table 2.5 (Continued)

5- Protection	6- Management	7- Monitoring
SCTA stated that "The involvement of merchants and owners, and the punctual restoration and revitalisation projects, are expected to set a new virtuous circle and to tackle the most significant threats to the property, reducing its vulnerability to negative developments that could affect its authenticity and integrity in the medium and long-term".		In the case of social indicators, all the data regarding the population residing and working, and the social activities within the nominated property and the buffer zone, will be collected by the Umdahs in coordination with the SMU. In the case of planning and development indicators, the regular check of the state of advancement of the projects being developed in the vicinity of the nominated property and buffer zone permits to verify the overall coherence of the rehabilitation plan for the Historic City with the on-going transformation of the central sector of Jeddah metropolis.

2.4 Building Information Modelling (BIM)

After the United Kingdom government made it mandatory for all public sector projects to adopt Building Information Modelling by 2016, key institutions in the construction industry, such as the Royal Institution of Chartered Surveyors (RICS), have published guidance documents with regards to the definition and application of BIM in the design process; there are changes in the associated legal outlines for projects adopting BIM and the impact of BIM adoption on the building sector. However, the story of BIM actually started in the 1970s, under the name of building product modelling (Eastman, 1975). Since 1995, when the International Alliance for Interoperability (IAI) started to develop a standard format in order to move between different software tenders used in engineering sectors, the issue of BIM and the necessity for it have become wider industry considerations (Kiviniemi, 2011). Furthermore, the development of BIM in the first ten years was mainly focused on the technology needed to deliver the concept of exchanging the data within the BIM workflow among different participants (Kiviniemi, 2008).

2.4.1   The concept of BIM

The first and basic concept of applying BIM (Building Information Modelling) is to design and build new buildings with high performance in each sector of the building industry, such as in terms of energy, economic and structural analysis, as well as the scheduling of the works as an addition to these new designs (Eastman et al., 2011). Nowadays, Building Information Modelling (BIM) is used in different contexts and for different purposes in the architecture, engineering, and construction industries (AEC). Thus, each sector of the engineering industry (AEC) has its own definition of BIM. Figure 2.7 shows the Building Information Modelling relation with engineering sectors.

BIM has been described by the NBIMS-US US National Building Information Model Standard Project Committee as "a digital representation of physical and functional characteristics of a facility", which is "a shared knowledge resource for information about a facility forming a reliable basis for decisions during its lifecycle; defined as existing from earliest conception to demolition" (NBIMS-US, 2015). Regarding this explanation of BIM, it is more than a 3-D visualisation; it

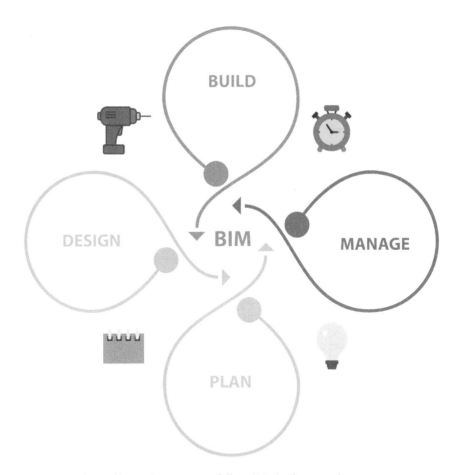

Figure 2.7 The Building Information Modelling (BIM) relation with engineering sectors

is designing, taking into account stakeholders, and managing it like an umbrella that encompasses different tools and platforms inside an organisation described as a BIM environment (Gupta, 2013), and this environment aims to achieve improved productivity and quality (NHBC, 2013).

Arayici et al. (2009, p. 1342) used the expression "BIM" to "describe a range of disciplines—specific software applications that support all phases of the project lifecycle from conceptual design and construction documentation, to coordination and construction, and throughout ongoing facility management, maintenance, and operations". Further, Murphy (2012), defined BIM as "the assembling of parametric objects which represent building components within a virtual environment and which are used to create or represent an entire building" and the objects "are described according to parameters, some of which are user defined and others, which relate to position in a 3-D environment relative to other shape objects". These objects, such as walls, windows, and doors, are linked to each other and have their own values (dimensions, materials, etc.) within the BIM databases. In each database, all the objects are linked to the whole building, and the relationship between the building's objects allows the database to update and modify selected objects at the same time and for all of the project members.

The BIM models are referred to as the Building Information Models. In this 3-D model of the project, the model can contain information regarding each element of the project, which can as also include the element's attributes. This information can be presented as geometrical, positioning coordinates associated with a project. For example, quantities, details regarding materials, cost estimates, project schedules, structural information, and data relating to energy. The users of the BIM can easily extract this information. Within the BIM, it is very easy to observe all the interrelation between the construction documents, resources, and the specifications of various items, the overall performance of the project, even before a project is built (Arayici and Aouad, 2010; Eastman et al., 2011; Gupta, 2013; Murphy, 2012).

BIM models are characterised by Gupta (2013), as "components that include data that describe their behaviour as needed for analyses and work processes; for example, quantity take off or energy analysis" and the models can contain "consistent and non-redundant data, such that changes to a component data are represented in all views of a component and the assemblies of which it is a part".

2.4.2   *The benefits of BIM*

BIM can offer a lot of benefits in terms of improving and supporting building projects undertaken by various individuals including owners, architects, designers, and construction engineers. In the case of owners' benefits, BIM can offer a better understanding of the project in different ways such as with regards to the concept, design benefits, and feasibility.

Through using the BIM tools, the building performance and quality can also be increased. Additionally, using a tool such as the IPD (Integrated Project Delivery) feature can improve collaboration among the project team, which allows the overall design and cost to be better understood. In the case of architects and designers, BIM can offer earlier and more accurate visualisation of the project. This would result in fewer corrections in a situation whereby changes have been made on the design. The BIM tools can also generate very high quality and accurate "2-D" drawings at any stage of the design process.

During the design stages, these models can extract accurate cost estimates. Additionally, the BIM models can be linked to energy analysis tools, which can improve the energy efficiency and sustainability of the project. In the case of the construction field, BIM can offer quick reactions to any design changes and highlight any errors and omissions before beginning the actual construction.

BIM can provide a better implementation of lean construction techniques. As for the pre-construction advantages, employing BIM in any project can improve the commissioning and handover of the FI, "facility information". By employing BIM in the project, it will result in better management and operation of facilities and allow for integration between both systems of management and operation (Eastman et al., 2011).

2.4.3   BIM for the project's lifecycle

The use of BIM in the building industry is a crossover between the design and planning boundaries which covers the whole lifecycle of a project. This provides support in terms of the procedures such as facility operation, construction management, project management, and cost management.

2.4.3.1   BIM in construction management

Working in the building industry sector is often difficult and provides a lot of challenges, such as delivering the project successfully, notwithstanding limited manpower, limited budgets, limited information, and accelerated schedules. In the building sector, sub-sectors such as architectural designs, structural designs, and MEP (mechanical, electrical, and plumbing) designs must be well manged and coordinated since it is not possible to undertake two tasks in one place and at the same time. Building Information Modelling (BIM) plays a significant role in detecting such clashes at an early stage. One of the main ideas of BIM is to envisage the realistic structure of a project before the actual physical structure is built, which can lead to improved safety, and a decrease in both the uncertainty and potential issues, as well as to simulate and analyse any possible impacts (Smith, 2007).

BIM can provide the chance for the sub-contractors to input any information into the BIM model before any on-site works begin, thus offering the possibility to pre-assemble or pre-fabricate some structures off-site (Smith, 2007). Using BIM can also provide the opportunity to easily extract the quantities needed and the material properties. Also, within the BIM application of the model visualisation, it can be easier to prevent errors through the use of the clash detection feature. These applications can highlight any clashes, errors, and overlaps, which can occur in any part of the project.

2.4.3.2   BIM in facility management "operation, maintenance, monitoring, and protection"

Facilities management (FM) can be described as the instruction frame for several operations, actions, and maintenance services in order to provide the main functions of an in-use project or facility (Yalcinkaya and Singh, 2014).

Employing BIM in any project can offer more transparency in relation to the handling of a project's information by the architect's team, the on-site engineering team, and the owners of these projects, or even the operators. This is done through offering each team the chance to add or modify aspects, with the ability to reference all information back to the first day the BIM model was created, which can lead to several advantages for both project operators and owners (Leite et al., 2011). For instance, if a project owner identifies any issues with a leak in the project, it is very easy to check the BIM model and to identify the location of the water valve, which can be better than investigating via the physical project. Moreover, within the BIM model, the owner can identify many of the building element properties, such as the size of a specific valve, the manufacturer, and other key information, which could not be examined as easily in the past (Leite et al., 2011).

The project dynamic information, such as the control signals and the sensor measurements from the project systems, can also be integrated into the BIM tools in the case of supporting the operation and maintenance analysis of the project (Liu and Akinci, 2009).

2.4.4   BIM maturity levels

To support the concept of using BIM in the United Kingdom and to establish what criteria are required, BIM has been defined within a range from 0 to 3 (see Figure 2.8). The main purpose of the BIM levels, according to Group (2011), is "to categorise types of technical and collaborative working to enable a concise description and understanding of the processes, tools, and techniques to be used".

The first level, which is described as level 0, is defined as the simple form of engineering data, such as the 2-D CAD and paper drawings.

The second level is level 1, which is defined as the use of 2-D and 3-D CAD in the format of the British Standard (BSI PAS1192, 2007, p. 1), this standard is according to BSI (2007) "the methodology for managing the production, distribution, and quality of construction information, including that generated by CAD systems, using a disciplined process for collaboration and a specified naming policy". At this level, sharing the data is managed by the provider and based on the CDE (Common Data Environment). Unfortunately, level 1 of BIM is the common operating level in many organisations; however, each discipline of the organisation publishes and maintains its own data without any collaboration.

The third level is level 2, which is distinguished by collaborative working, also, this level is described as the "3D Environment held in individual discipline 'BIM' tools with attached data" (Zheng, 2013). At this level, the most important thing is how the information can be exchanged between various participants in the project by using common CAD and document formats, such as schedules, emails, drawings, certificates, etc., which according to NBS (2015), allows "any

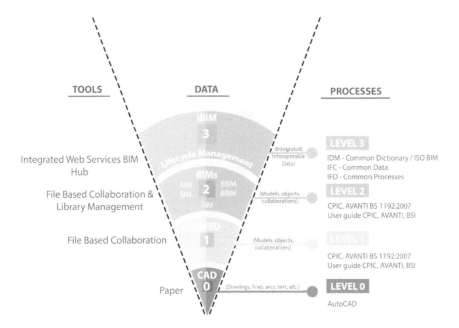

Figure 2.8 BIM Maturity Diagram Mode

organisation to be able to combine that data with their own in order to make a federated BIM model, and to carry out interrogative checks on it".

According to the RICS (Royal Institution of Chartered Surveyors), one of the significant standards of level 2 BIM is "that as a project progresses and information grows, it moves from information fit for design, to information fit for construction and then to information that represents what has been constructed". The outcome of level 2 BIM is "therefore an accurate record of project information", which can be used to support the built asset in operation.

The fourth level, level 3, as defined by Boon and Prigg (2012), is "the fully open process and data integration enabled by web services compliant with the emerging IFC/IFD standards, managed by a collaborative model server". This integrated BIM (IBIM) can employ concurrent engineering methods (Group, 2011; NBS, 2015).

2.4.5 BIM modelling software

The main concept of BIM software is that it offers multiple models lying on top of one another for several purposes, such as to identify how these models fit together to analysis and manage them and to share information. The current available BIM software on the market is proprietary, owned and developed by software vendors. There are many types of BIM software in the engineering market, however, two of the leading pieces of BIM software being Revit from Autodesk (Autodesk Inc, 2015) and ArchiCAD from Graphisoft (Graphisoft SE, 2015). Table 2.6 outlines some of both the strengths and weaknesses of pieces of the software.

Table 2.6

BIM Modeling Softwares

Autodesk **Revit**   Graphisoft **ArchiCAD**

Autodesk **Revit**	Graphisoft **ArchiCAD**
Architectural design applications and are based upon the idea of BIM	
Strong intuitive design tool and it has a friendly interface for the user	The tools are well designed and the software interface is friendly for the user
Containing huge object libraries (which were developed by Autodesk and third parties "featuring 13,750 product lines from 850 different companies")	The tools are well designed and the software interface is friendly for the user
Autodesk Revit deals with the custom objects in a spatial environment called Revit Family "RFA" Great feature in Revit Family is that the user can create a "mass model", containing a combination of void and solid forms, and the mass façades can be turned into	It supports the user in creating custom parametric objects through a spatial environment which uses Geometric Description Language (GDL)
Revit API (Application Programming Interface) offers great support for any external application development	Very strong in supporting the design, building, systems and facilities management of applications
Weak when dealing with large files	Weak in terms of its capabilities in relation to custom parametric modelling
Limitations when dealing with the laser scanning point cloud	Weak in terms of the in system memory always affects large files' processing

2.4.6   The integration within BIM

In the case of BIM, data integration refers to the process of connecting different formats of project information into one file format. This file format is suitable for working within multiple platforms. It has been noted that the interoperability between the same vendor's applications, such as Autodesk, can be done fluently. On the other hand, a huge number of stakeholder's programs are needed in order to have a common platform for the design and the analysis of the structures by different enterprises. Many of the companies involved in the building industry often work with program applications that are determined by their operability and their cost. This can cause many complications and difficulties in cases when transferring data among the different departments involved in the construction processes of the project (Darie, 2014).

A number of research studies have been conducted on pre-data processing and the integration of information in BIM for heritage buildings. However, exchanging and integrating the data face causes a number of issues in terms of digital-born material moving into the BIM environment (e.g. the Green Spider plug-in for Autodesk Revit 2012, as shown in Garagnani and Manferdini [2013]). The issue of post-data processing for attributes to be improved and integrated into different datasets remains unresolved (Saygi et al., 2013). One of the proposed solutions for fixing this issue is using a standard format, such as the COBie standard, which is currently being applied in the UK. The COBie standard, which is an abbreviation for the Construction Operations Building Information Exchange, can be described as a spreadsheet data format that includes all the useful information regarding the building in digital format (BIM Task Group, 2013). Another standard is the IFC "Industry Foundation Classes", which was developed by the international Alliance for Interoperability; the main purpose of which is to cover both geometries and semantics (Kolbe et al., 2008).

In recent years, a number of new ideas have been offered in relation to both the commercial and open source software solutions for post-data processing and data management. Some of these solutions can be found in the Building Information Model-server, depending on the format of IFC. The Open BIM movement introduced via Graphisoft, Tekla, and building-SMART, aim to improve the quality of building data that is related to elements or even complete buildings. On the other hand, improving the attribute management has not been an important aspect of the solutions yet. Consequently, the external attribute data integration has not been fully discovered.

The heritage architectural object requires an integrated representation within several thematic drawings. This can be described as the multi-layered modelling approach; which can include themes such as alterations, structural systems, and material usage and can be represented in a multi-layered 3-D environment. Even though BIM offers temporal representations (4D), some other thematic representations are beyond the capabilities of today's Building Information Modelling approaches (Saygi et al., 2013).

2.5   The evolution of the usage of Heritage Building Information Modelling (HBIM)

In the last 20 years, to conduct any inventory work and take measurements for a building, a surveyor and his equipment, such as a total station, has been the first choice for accomplishing this task (Santana-Quintero et al., 2008). The surveyor's work usually starts with the total station in order to capture the key points of the site and the façade of the building. Afterwards, these key points are input into a CAD programme back in the office, after which the surveyors draw up the plans and elevations of the site or the building in 2-D. On the negative side, the generated data via this method is not always completely accurate, it can be slow, and it can include many errors and lack detail. It is also hard to visualise 2-D plans in 3-D (eBIM, 2015).

These traditional surveying methods using scanning do not currently result in a product that is ideal for the purpose of the BIM process due to the fact that the heritage sites use non-parametric CAD programs in order to generate plans for survey proposals (Thomson and Boehm, 2015).

Employing BIM into the historical field has been introduced as a new method over the past few years. This can be considered as the fundamental stage toward the use of BIM workflow for retrofit and reconstruction tasks (eBIM, 2015). This new method is described as Heritage BIM (HBIM). It is occasionally also described as Historic BIM (HBIM), a somewhat narrower term (Murphy et al., 2009).

This method has been used in a number of projects worldwide such as in the research of (Dirix, 2015; S. Fai et al., 2011a, 2011b; Hichri et al., 2013b; Murphy et al., 2009; Murphy, 2012; Oreni, 2013; Penttilä et al., 2007). The main themes of these pieces of research and projects were to provide intelligent data (S. Fai et al., 2011a, 2011b), and "an as-built" digital 3-D model (Backes et al., 2014), to be used for several purposes such as documentation, conservation, and management.

Almost all of these projects employed terrestrial laser scanning (TLS) and close-range photogrammetry as the fundamental step for data capturing. This step is commonly referred to as Scan to BIM or in heritage cases Scan to HBIM. Theoretically Scan to BIM/HBIM as an expression is incorrect as the end result is not Building Information Modelling (BIM) as commonly known, in other words "the process", yet, in reality, is a 3-D parametric object model that aids the procedure at its current "LoD" level of development (Thomson and Boehm, 2015). While, from a BIM/HBIM viewpoint, generating parametric 3-D models of a building or site from scan data is a recent development, it is actually not particularly new. The beginning of commercialised terrestrial laser scanning systems lies in generating parameterised surface representations from segmented point clouds (Runne et al., 2001). There are even earlier references than this in the aeronautical area for parametric surface reconstruction (Haala and Anders, 1996, pp. 285–290; Runne et al., 2001).

Additionally, Murphy (2012, p. 13) described HBIM as "a novel prototype library of parametric objects, based on historic architectural data", as well as "a mapping system for modelling the objects library", based on the survey data of terrestrial laser scanning and image survey data.

Penttilä et al. (2007) employed BIM to "demonstrate how modern digital methods can be used in reconstruction design" and in addition, "in renovation projects, with special emphasis on BIM". In this study, the idea of an "inventory model" was employed on an existing building (which was the building of HUT/Architectural Department in Finland), representing an important basis for the inventory data being well structured, presenting both the present and past situations of the building. It can be noted that in Penttilä et al.'s (2007) study that the inventory model idea covered the definition of historic data and how it can be implemented, however, using BIM as a database for documentation has not yet been totally investigated.

Fai et al. (2011a, 2011b) employed BIM into the heritage field for the purposes of documentation and conservation based on the terrestrial laser scanned survey data in Toronto, Canada. Murphy (2012) applied BIM to classical architecture in Dublin, for recording and documenting purposes, during the period from 1700 to 1830, by using "historic architectural data", as well as "a mapping system for creating the library objects onto laser scan survey data".

Furthermore, the research in HBIM can be organised in groups regarding the institute or the organisation from which it came. The next sections will give more comprehensive descriptions about the approaches of other researchers such as DIT in Dublin, CIMS in Carleton, Polytechnic of Milan, CNRS, and FBK in Trento.

2.5.1   Heritage BIM worldwide

2.5.1.1   DIT—Dublin Institute of Technology (Maurice Murphy et al.)

The DIT employed BIM for use in the heritage field for recording and documenting purposes in Dublin, Ireland. They described this process as Historic Building Information Modelling (HBIM).

The HBIM process consists of the survey of existing structures via remote collection through using the terrestrial laser scanner survey data. This is then followed by point cloud processing and generating an ortho-image model (Murphy et al., 2013) (see Figure 2.9). The next step of Murphy's method is to create a parametric library based on the laser scanning survey data, using the architectural shape rules and shape grammars from the "18th century architectural pattern books". Murphy's library was created by employing the Geometric Descriptive Language (GDL) for the ArchiCAD BIM program. In the final step of the HBIM, Murphy claimed that the "final HBIM" outcome was a "product consisting of 3-D models of the building, including the detail behind the object's surface, relating to its methods of construction and material makeup".

The resultant HBIM can then be used for automatically producing conservation documentation and analysis of historic structures in addition to visualisation. The systems used for mapping or (modelling) the objects into point clouds or mesh models are based on manually aligning the objects onto orthographic projects from the point cloud or mesh model.

More recent work of DIT is concentrating on automation for improving the current slow process of converting unstructured point clouds into structured semantic BIM components. The principal results from the research, which converts laser scan or photogrammetric survey data to HBIM, is the creation of intelligent computer models. These computer models can be used for conservation analysis and solutions for historic structures and their environments. The next stage of DIT work is to convert the HBIM model into CityGML for further GIS

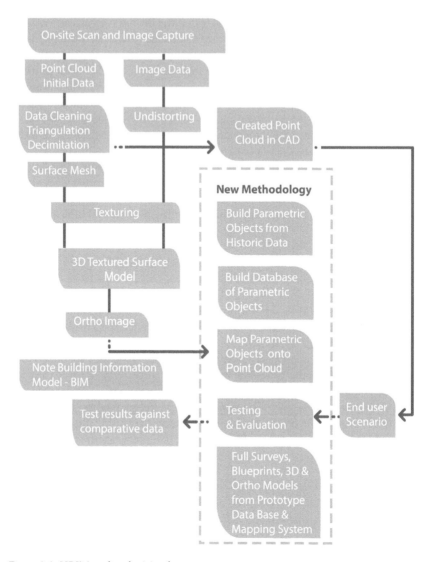

Figure 2.9 HBIM outline by Murphy

(geographic information system) analysis. For this purpose, Google SketchUp with the CityGML plug-in is used. Via using the CityGML (an open data model and XML-based format for the storage and exchange of virtual 3-D city models), it is possible to provide further capabilities for linking the information systems with 3-D heritage models. Moreover, the CityGML model can be integrated into GIS systems for analysis and efficient management, which is required for maintaining important heritage monuments (Dore and Murphy, 2013).

2.5.1.2 CIMS—Carleton Immersive Media Studio (Stephen Fai et al.)

The CIMS employed BIM into the heritage field for the purposes of documentation and conservation based on the terrestrial laser scanning (TLS) survey data in Toronto, Canada. The HBIM model of CIMS was developed by using the available software packages that are appropriate for specific applications such as AutoCAD, Civil 3-D, SketchUp, Revit (Fai and Rafeiro, 2014). Furthermore, S. Fai et al. (2011a) focused on the issues related to linking the laser scanning data within BIM, and how to model generic objects like library objects based on laser scanning data within a BIM environment. The hybrid documentation model, which was offered by Sabry El-Hakim (2005) and Fabio Remondino (2003), was used to develop the case study model of this project in Toronto. The CIMS focused on other issues related to the levels of detail (LoD) and they tried to identify the suitable LoD for these purposes.

The CIMS HBIM model contained different types of datasets, such as building type, performance, construction, and materials, with regards to a "digital object" model that could permit access to and on-going authentication, based on the separate assets that make up the whole (S. Fai et al., 2011b). Moreover, their case model "is not wholly parametric, it points to the potential of parametric relationships between all data types for heritage documentation".

The 3-D environment was created to allow navigation for maintaining the BIM intelligence of the objects that constitute the model. By doing so, it is possible to examine the model at both micro and macro scales with no loss of reliability or data. Then, intangible heritage information was incorporated such as historical images and texts, multi-language storytelling, and music. Finally, the HBIM model can be viewed trans-temporally, by employing the timeline function of the project management software. Furthermore, the CIMS developed a BIM database, which is the CDMICA "Cultural Diversity and Material Imagination in Canadian Architecture". This database can be used as a tool for heritage sites and shows the results of the East Façade (LoD 3) of the BIM with point cloud overlay (Fai and Rafeiro, 2014).

2.5.1.3 ABC Department, Polytechnic of Milan (Daniela Oreni et al.)

The main aim of the ABC Department is to use the HBIM model to provide a valid support to understand and interpret the structure (e.g. elements like cracks or other structural damages can be easily inspected). In addition to the aforementioned aim, the HBIM should overcome the lack of parametric model software (i.e. Graphisoft, ArchiCAD, and Autodesk Revit) for the management of complex and irregular shapes, as well as the issues concerning the standardisation of elements in objects and families. The surveying work was based on total stations and the laser scanning survey. Image-based methods were used to reconstruct areas where laser data did not provide a sufficient level of detail or where there was a complete lack of data (Oreni et al., 2013).

To start with the HBIM modelling, the 3-D model was divided into its structural elements following the constructive logic of the building: i.e. vaults, wooden elements of the cover, walls, columns, stone ashlars, and decorative elements. The software used were Rhinoceros and Bentley Pointools. Starting from multiple point clouds, Rhinoceros allowed the representation of complex shapes by using the Boolean operations (i.e. simple extrusions and revolutions on generative axes). NURBS (Non-Uniform Rational Basis-Splines) was employed to generate surfaces that arise from a spatial deformation of a grid of square-tiled surfaces (patch surfaces) in which the positions of the control points determine the deformation. The second step was to use the model and its elements in BIM software, maintaining the parametric data for each shape along with the possibility to add information regarding materials, the state of conservation, the planned intervention, and so on. Finally, a rigorous conversion procedure was developed in order to obtain a complete parametric conversion of all the elements without losing information.

In the ABC Department project, the interoperability between Rhino and Midas (a software for finite element analysis—FEM) was tested in order to use the accurate model and to evaluate project solutions for the new structures and the consolidation of the case study (D. Oreni et al., 2014). Furthermore, in recent projects of the ABC Department, Autodesk Revit has been used for the BIM process, and specially for modelling the simple objects. Moreover, ABC Department produced a detailed HBIM model which was then converted into a model for mobile apps based on augmented reality (AR) and virtual reality (VR). The aim of this model was to extract useful information not only for expert operators, but also for a wider user community interested in cultural tourism (Barazzetti et al., 2015).

2.5.1.4   *3DOM-FBK Trento, 3D Optical Metrology Group—Bruno Kessler Foundation (Fabio Remondino et al.)*

The aim of the 3DOM-FBK project is to provide 3-D documentation and visualisation of the archaeological building based on terrestrial laser scanning (TLS) and image-based techniques. The early project of the 3DOM-FBK involved developing a 3-D mesh model of the heritage building. In this model both TLS and image-based datasets have been merged using Innovmetric Polyworks (Remondino et al., 2009). In 2013, the group focused, tested, and gathered information on some of the capabilities of BIM and GIS through workflows for 3-D modelling and information management of historic buildings prior to the decision-making process of conservation. The HBIM model was done via Autodesk Revit Architecture 2013 (Saygi and Remondino, 2013). Recent works of 3DOM-FBK were focusing on the 3-D textured polygonal models based on the terrestrial photogrammetry for large-scale projects (Remondino et al., 2016).

2.5.1.5   *CNRS—Centre National de la Recherché Scientifique (Livio De Luca et al.)*

The main focus of CNRS is on photogrammetry and data acquisition (De Luca et al., 2014). However, the group has interest in developing a semantically enriched BIM model (Hichri et al., 2013a). The process starts with a point cloud and leads to the well-structured final BIM; it also proposes an analysis and a survey of existing approaches on the topics of acquisition, segmentation, and BIM creation. Furthermore, it presents a critical analysis of the application of this chain in the field of cultural heritage.

The developed approach of CNRS suggests linking the first step of acquisition and the final "as-built" BIM. Semantic features will be affected by historic objects directly in the survey and the segmentation stages on the basis IFC classes.

As claimed by the group, in the next years this approach will be implemented by creating a communication platform between common laser scanner software and BIM software (Autodesk Revit and Faro Scene). Besides, this communication will be ensured by a common database following the IFC classification model (Hichri et al., 2013a).

2.5.2   BIM and the heritage field

BIM can contribute to the heritage conservation field in several ways, for instance, in terms of developing the understanding of historic buildings and their context, knowledge about materials, construction techniques and the building's pathologies; while also understanding that heritage buildings cover a wide range of materials and assemblies that are not documented and are not presented within the stock libraries of 3-D model objects. Furthermore, BIM can provide lots of advantages to the historical field, for example, providing a built digital information model about the heritage building (eBIM, 2015), providing a full study of any future restorations and changes before final decisions are made, supporting the building's maintenance, aiding with budgeting for the purposes of repairs, maintenance, and besides, offering a wider public building experience as models can be viewed with free viewer software from remote locations (Logothetis et al., 2015). According to Fai et al. (2011b), "HBIM can provide automated conservation documentation and differs from the sophisticated 3-D models produced from procedural and other parametric modelling approaches, whereby the main product is a visualisation tool". Also, employing BIM in the heritage field can provide technologies that can answer the increasing demand for multidisciplinary knowledge and in the world of Cultural Heritage, there is understanding that BIM and WEB-BIM procedures can effectively contribute to the management of historical and architectonical heritage data as has been marked in numbers of studies (e.g. Fregonese et al., 2015; Garagnani and Manferdini, 2013; Lipp et al., 2008).

Table 2.7 shows the main differences between BIM and HBIM.

2.5.3   HBIM challenges

There appear to be many challenges that can be faced by those using BIM in the historical field; these challenges can include organisation, technical issues, and problems with the site.

The organisational issues can be related to the cost to the employers of utilising new methods, as well as the cost of the new technology and professional training required. The technical issues can include such things as the high cost of the BIM

Table 2.7

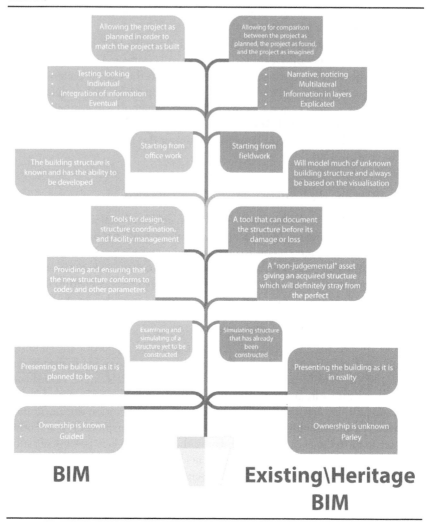

applications and errors in the system and training time. Besides that, compatibility issues can arise with files and the computer platform.

One of the most important challenges is related to the modelling difficulties (Fregonese et al., 2015). Indeed, Architectural, Engineering and Construction (AEC) modelling applications such as Autodesk Revit and ArchiCAD are not commonly used to model heritage buildings (eBIM, 2015).

Regarding the historic site issues, one of the significant points to be considered is how complicated they are, which can affect the cost and how long it takes to undertake the survey and also, the accuracy of the survey can be problematic. A huge file can be generated from modelling at such complicated sites (eBIM, 2015).

A lot of the historical sites have unique architectural "non-standard" elements, and each of these unique elements and parts needs to be modelled up, such as walls, structures, windows, doors, etc. (Fregonese et al., 2015; Saygi et al., 2013). On the other hand, using BIM for a new project has the advantage of providing available libraries of the modern building elements, which are available with the BIM applications, or even through online 3-D libraries (Garagnani and Manferdini, 2013).

Other important obstacles are access and permission issues, which can occur when attempting to undertake recording of these historical sites. All these issues can possibly increase the costs of the project.

2.5.4   Modelling in HBIM

The most important aspect of HBIM is transferring the information, which is based on the rich data survey, into 3-D parametric modelling. Since the BIM has been employed in the heritage field, almost all research studies and projects use manual methods for 3-D modelling, while very few use the semi-automatic methods, especially to model the mechanical, electrical, and plumbing (MEP) parts. The need for better 3-D documentation of the built environment has arisen during the last decade, driven in most parts by city modelling on a large scale and BIM on a smaller scale. Automation is seen as desirable as it is less time-consuming and therefore, reduces the amount of costly human intervention in the model creation procedure. Currently, traditional surveying does not produce ideal outcomes to be used in the BIM process. This is due to the historical use of non-parametric CAD programmes in order to generate the survey 2-D drawings. Consequently, a paradigm shift is required in the work processes and displaying techniques of the stakeholders, who are undertaking this work in adjusting to this shift (Thomson and Boehm, 2015).

In the case of the HBIM modelling process, the modelling represents the HBIM's objects construction, which represent the heritage building components and parts, containing both non-geometric and geometric features and relationships. And according to Volk et al. (2014, p. 109), if HBIM "is modelled on the basis of previously captured building information, the preceding data capture, processing, and recognition methods influences data quality through the deployed technique and the provided LoD". The key point of comparison between the modelling methods is related to the modelling accuracy, or in HBIM's case, it is known as LoD "level of details/development" (Tang et al., 2010; Volk et al., 2014). However, there is no standard for HBIM in the case of the LoD. Some tried to establish their own standards as "as-built" BIM modelling (section 2.5.5 will discuss more about this). The main issue with the "as-built" BIM level of detail is the huge amount of details that are captured via advanced technologies, such as laser scanning, which can require lots of time and cost more. In this case, it is very important to develop automated, or even semi-automated, modelling methods to address this matter.

By focusing on a number of projects, automated BIM/HBIM modelling or changes of the surface models toward volumetric, semantically rich substances is in its early stages (Arayici, 2008; Xiong et al., 2013). Moreover, several studies have dealt with semi-automated modelling in relation to building the surfaces or components as geometrical representations. On the other hand, these studies did not devote attention to the component properties or semantic information (Arayici, 2008; Dai et al., 2010; Frahm et al., 2010; Klein et al., 2012; Xiong et al., 2013). The integration of the non-geometric features, such as the relational, economical, functional, or semantic information of the heritage building, has always been done via interactively or semi-automated methods (Donath et al., 2010; Hajian and Becerik-Gerber, 2010; Thomson and Boehm, 2015; Xiong et al., 2013).

2.5.4.1   Manual, semi-automated, and automated approaches

MANUAL

In general, the 3-D digital modelling takes place in order to represent or simulate a structure that does not exist yet. On the other hand, for existing buildings, such as heritage buildings, the aim is to provide a model of these structures as they are in reality. Presently the procedure is more likely to be manual and it is accepted that it is time-consuming, uninteresting, subjective, and requires more skills by many experts and research studies (Rajala and Penttilä, 2006; Tang et al., 2010; Thomson and Boehm, 2015). The manual approach can be described as the generation of 2-D CAD plans, out of possession of the LIDAR data demands of the user in order to employ the LIDAR data as an instructor in the BIM applications. This results in successfully being able to shape the geometry, demanding an elevated knowledge in order to understand the scene, in addition to increasing the rich semantic information which really promotes BIM as a useful procedure (Thomson and Boehm, 2015).

SEMI-AUTOMATED AND AUTOMATED

Both semi-automated and automated modelling are seen as necessary commercially in order to reduce the modelling time, as well as the cost, besides making the scanning a more sustainable scheme for the purposes of a wide range of assignments in the project lifecycle, for example, detecting the daily construction changes (Eastman et al., 2011; Thomson and Boehm, 2015). Transformations of the point cloud surfaces models toward volumetric objects, which can be described as automated BIM modelling, are still in their infancy (Arayici, 2008; Xiong et al., 2013). A range of researchers have presented a semi-automated modelling approach to the surfaces of the building or components with regards to their geometrical representations. On the other hand, they have not thus far considered the component characteristics or semantic information (Adan and Huber, 2011; Arayici, 2008; Eastman et al., 2011; Frahm et al., 2010).

In the event that the non-geometric characteristics, such as relational, functional, semantic, or economical information of the heritage buildings are combined with BIM, this is achieved via interactive or semi-automated methods (Donath et al., 2010; Hajian and Becerik-Gerber, 2010; Xiong et al., 2013). For example, to model covered components, such as pipes, ducts, plumbing, or conduits "HVAC/MEP", they can only be modelled through high user input at present (Dickinson et al., 2009).

2.5.5 As-built level for HBIM

One of the most important challenges for HBIM is determining suitable levels of detail or the levels of development. This is due to the fact that there are no international BIM standards currently available regarding the level of detail or developments in the heritage buildings field. The main reason for that, as Fai and Rafeiro (2014) stated, is that "we must reconcile the LoD of the BIM with that used in the documentation process" for both the terrestrial laser scanning "TLS", as well as the close-range photogrammetric survey.

In the case of the historical buildings, it seems that it is very important to know the main purpose of using BIM; and once we know that, we can determine which level the project should reach. Large numbers of historical buildings and sites are somewhere between conservation and preservation projects. Therefore, in relation to the conventions concerning protection, based on the World's Cultural and Natural Heritage sites, these historical locations must be conserved or re-built as they were. As a consequence, the LoD of these projects must be at the level of the as built/level 7 of the UK standards. The level of "as-built" BIM for heritage building is used so as to represent the physical and functional features of these buildings within BIM representation. This is particularly in terms of the state of the historic building at the time of surveying in order to obtain a semantically enriched model.

The "as-built" level of detail includes three aspects. The first aspect is the geometrical modelling of the object. The second aspect is the attribution of categories, as well as the material properties of the objects. The third aspect is the creation of the relationships to link these categories, materials, and the model (Hichri et al., 2013b). With regards to the huge amount of detail, which can be received via the laser scanning process, the procedure of the modelling is mostly manual. The characterisation of "as-built" HBIM, as Hichri et al. (2013a) stated, "involves three aspects, allowing the building of a structured point cloud: shapes, relationships, and attributes".

The first aspect, which represents the shape belonging to the object, can be classified into three dimensions (Tang et al., 2010). The first dimension is whether it is parametric or non-parametric. The second dimension is whether it is global or local. Finally, the third dimension is whether it is explicit or implicit. In the first dimension, in terms of the shape of the object being parametric, the object is defined by using a set of parameters: length, height radius, and so on (Campbell and Flynn, 2001). On the other hand, non-parametric objects use other ways of characterisation, such as the triangular mesh (Tang et al., 2010).

In the second dimension, in order to represent the object as global, the whole object is defined; however, to represent it as local, just one part of the object is characterised. An example of global representation can be found in the non-parametric representation, such as triangle meshes, by representing the whole object. However, the local representation can be found in such parametric representations (Tang et al., 2010). In the third dimension, to represent the object as an explicit or implicit representation, as Tang et al. (2010, p. 829) described; "Explicit representation permits direct encoding of the shape of the object", such as the triangular meshes; however, "Implicit representation allows an indirect encoding for the shape of the object, using an intermediate representation", such as the histogram for normal surfaces.

The second aspect, which involves representing relationships between objects, is key for the very important job of describing the positions and displacements of components in the case of BIM (Cantzler, 2003; Nüchter and Hertzberg, 2008) and HBIM. The spatial relationships in BIM and HBIM can be described as aggregation, topological, and directional relationships (Tang et al., 2010). The hierarchical-based tree representation can therefore be used to model the aggregation and to define the structure in a local-to-global way (Fitzgibbon et al., 1997). Both topological and directional relationships can be represented by a graph. The third aspect is related to what the object's attributes represent. Indeed, it is very important in a BIM and HBIM context to have this feature. Moreover, this feature can enrich the final 3-D representation by allowing characterisation of these objects, with information pertaining to features such as the materials, the state of preservation, and information about the documentation in relation to the heritage building, for instance, and also whether the object has been restored or replaced (Tang et al., 2010).

Hichri et al. (2013a) stated that "Attributes or object classes can be: graphical or alphanumerical". The difference between them is that the graphical attributes contain the required properties for the 3-D modelling, such as numerical values and Cartesian points, however, the alphanumerical attributes contain all the additional information, such as the composition, dimension, and economic data. Additionally, Hichri et al. (2013a) stated that "Attributes are also structured on a set of classes", and each object "is characterised by semantic information defining it". These classes can be "tangible", such as the floor, wall, and ceiling, or "abstract", such as the manufacturing procedure, cost, and relationships between classes.

2.5.6   Data acquisition in HBIM

2.5.6.1   Terrestrial Laser Scanning (TLS) in HBIM

During the 2000s and since the technology has become more commercialised, and one of the most popular technologies used to capture the 3-D complex structures to be modelled in the 3-D environment is terrestrial laser scanning (Budroni and Boehm, 2010). Figure 2.10 shows an example of a laser scanning device.

There are several reasons to use terrestrial laser scanning in Heritage Building Information Modelling and these relate to the high level of accuracy that laser

Figure 2.10 Figure C10 scanner

scanning can offer, in addition to the time necessary to do the scans on the cultural sites. Besides, the most important reason to use TLS is that it can be used to measure the existing condition of buildings (Cheng and Jin, 2006; Macher et al., 2014). TLS can be described as an automatic measurement scheme that measures the 3-D coordinates based on the surface from the selected object. The generated data that is received from the laser scanning is then represented in point cloud form. Each of these points, according to Murphy (2012, p. 22), has the "x, y and z coordinates" based on the scanned surface; also, the "laser ranger is directed towards an object by reflective surfaces that are encoded so that their angular orientation can be determined for each range measurement". These point clouds are described by Thomson and Boehm (2015) as the absolute minimal level of detail base (stylised as LoD 0) from which more data-rich abstractions can be produced to represent higher levels of detail.

Several laser scanning systems are available in the engineering market these days. On the other hand, as Murphy (2013, p. 22) proposes, "there are three types of scanners suitable for metric surveys for cultural heritage", which are "triangulation, time of flight scanners, and phase comparison".

The difference between these systems concerns the technique of how the scanner calculates the 3-D coordinate measurements. For instance, in the case of the triangulation type, the scanner uses the spot regarding the laser ray on the surface of the object, captured via one or more cameras (Murphy, 2013). In contrast, as Boehler et al. (2003) stated, "time of flight scanners calculate the range, or distance, from the time taken for a laser pulse to travel from its source to an object and be reflected back to a receiving detector".

Since 2009, the use of terrestrial laser scanning has been introduced as a common method for capturing the reality of the as-built BIM projects (Arayici et al., 2009; Hichri et al., 2013a; Huber et al., 2010; Volk et al., 2014). This was due to two main reasons (Backes et al., 2014); firstly, the endorsement of governments such as the UK and USA to focus on the issue of BIM, which was reflected in the increased use of laser scanning to capture reality (BIM Task Group, 2013), and secondly, the support of CAD vendors for the integration of point-cloud handling devices without plug-ins in their BIM tools, which resulted in a high demand for laser scanning.

Figure 2.11 shows the typical laser scanning survey method.

2.5.6.2 *Combining digital images and laser scanning*

The new modern laser-scanning systems usually incorporate built-in cameras for capturing the photos, then the photos are applied to the point cloud to be coloured via linking the multi-image batch to the point cloud data.

As stated by Abmayr et al. (2005), "the RGB colour data from the images can be mapped onto range data by taking account of point translation, instrument rotation, and perspective projection". For this, both the laser, as well as the camera, as pointed out by Murphy (2013), "must be correctly geometrically calibrated" and "the correction of the camera is presented to correct the distortion of camera

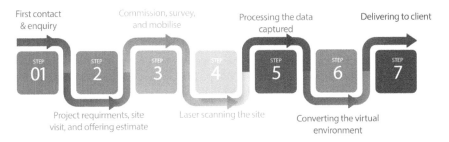

Figure 2.11 The typical laser scanning survey method
Source: Author

lenses, and by mapping onto the point cloud any perspective contained in the images is removed".

Additionally, the HDR colour images "can be precisely mapped onto a geometric model represented by a point cloud, provided that the camera position and orientation are known in the coordinate system of the geometric model" (Beraldin, 2004, p. 232).

2.5.6.3   *Terrestrial laser scanning data processing*

Laser scanning usually captures a huge range of real-time data, which is represented in 3-D coordinates, described as 'point cloud data', always collected from several scan locations. This is because it is usually impossible to scan the whole object in one scan. In that case, it is very important to register the different scan's data in order to locate all the 3-D point cloud data together via coordinate transformation (Cheng and Jin, 2006). While laser scanners can take a few minutes to scan millions of accurate 3-D points, there is enormous work involved in transporting this data into a 3-D model containing useable information. Moreover, as Remondino (2003) explains, "dedicated software programs such as Leica cloudworks, Polyworks, AutoCAD, and RiScanpro have highly improved the processing, manipulation and analysis of vector and image data from the point cloud" and "all of these software platforms have combined algorithms for triangulation and surfacing of the point cloud".

DATA CLEANING

Once the scanner's point cloud is transferred, there are several suitable programs that can be used in the step that involves the evaluation and filtering of the point cloud data. The data evaluation can be described as a process of locating the overshoot and repeated zone when scanning and deciding whether to scan over. The data filtering can be described as reducing the noise, smoothing noise, and resampling the data (Cheng and Jin, 2006).

Each system from these laser scanners has its own software. For instance, Leica Cyclone deals with the Leica laser scanning systems. Generally, the modelling process in Leica Cyclone involves generating the best-fit geometric objects out from the point cloud. Leica Cyclone has a number of object-suitable utilities for the user to select from, depending on the topology that comes out of the scanned point cloud. However, as Ikeuchi (2001, p. 119) states, "other applications are used instead of polygonal 3-D models; NURBS surface models, or editable feature-based CAD models".

The concept of point cloud registration is to combine a range of scan stations with different perspectives of the scanned object. According to Leica Geosystems (2006), "integration is derived by a system of constraints, which are pairs of equivalent tie-points or overlapping point clouds that exist in two Scan-Worlds". The registration method "computes the optimal overall alignment transformations for each component Scan-World in the registration such that the constraints are matched as closely as possible".

There are several methods which can be used to register these scan stations by applying one of the two methods, or a combination of them, for example, target based or point cloud based (Mills and Barber, 2004; Rajendra et al., 2014). Using the Global Positioning System (GPS), the coordinates based on the laser scanner location can be determined, which can then, according to Cheok et al. (2000, p. 466), "allow for the scans from each position to be brought into a common frame of reference in a global or project coordinate system".

The first method, target based or target-to-target registration, is a registration between multiple point cloud scans and registered into a single point cloud through the use of control targets in the point clouds. Moreover, to achieve perfection and with minimum errors, three corresponding points, or more, at each point cloud registration must be in common between them. These points can be either natural or artificial targets. This technique is designed to offer the most accuracy inasmuch as the standards of survey net plan are respected (Pfeifer and Böhm, 2008). Despite natural targets being assigned manually, the artificial targets can be assigned automatically via employing certain algorithms. The automatic assigning procedure is subject to detecting certain shapes from targets, such as spheres or HDS black-and-white "B/W" targets.

The second method is the overlap or the cloud-to-cloud registration method, which involves aligning overlapping scans into a single point cloud through using a number of constraints that are nominated within the point cloud software, such as Autodesk Recap Pro or Leica Cyclone model space. It is very important in this method that the selected features are matching in both registered model-spaces (Darie, 2014). The registration result can be used as a base frame of the geometry of the historic structure as in the research of Murphy (2012), which can then be modelled in different modelling software, such as Autodesk Revit, 3-D MAX, SketchUp, and AutoCAD.

2.5.6.4 Architectural photogrammetry

In some circumstances, the terrestrial laser scanning has some limitations, especially in the case of complicated sites and difficult areas being reached and further, for detailed structures and vignettes. Architectural photogrammetry or the close-range photogrammetry technique can offer a good solution for such issues, which can produce both orthoimage and linear drawings (Cheng and Jin, 2006). As Hanke and Grussenmeyer (2002) stated, architectural photogrammetry "is a technique for gathering information of the obtaining geometric, e.g. size, position, and shape of every object, which was imaged on photos before". Recently, architectural photogrammetry and 3-D modelling methods have been rapidly advancing, with part of the focus on developing new technologies for the purpose of applying them to architectural heritage documentation. Current advances have allowed for semi-automated and automated solutions, based on Dense Stereo Matching (DSM) (Furukawa and Ponce, 2010; Hirschmuller, 2005) and Structure from Motion (SFM) (Agarwal et al., 2011, pp. 105–112; Pollefeys et al., 2008; Vergauwen and Van Gool, 2006), to be presented commercially. These methods have an advantage over the web-based software through providing accurate results in order to obtain 3-D point clouds, as well as the textured mesh surfaces, such as Autodesk Recap® and PhotoModeler by Eos Systems Inc.

The full automation of the procedure of image orientation and matching has been shown to facilitate and speed up the data processing task that, in some cases, could be a never-ending process if performed manually. The dense cloud of 3-D points as a result, by itself, is typically not very useful. In this regard, a number of post-processing steps are usually carried out. This post processing often involves filling holes, filtering, smoothing, meshing/triangulation, and sometimes NURBS patch conversion. A separate 3-D software program generally carries out these functions (Alitany et al., 2013a).

2.5.7 Level of accuracy

Laser scanning is an advanced technology which can provide highly accurate data. According to Murphy (2012) using laser scanning and photogrammetry in the case of capturing the heritage structure can "meet with the accuracy and efficiency requirements for recording and surveying of historic structures and artefacts". However, accuracy can be affected by various factors, for example errors in the distance and the angle measurements, and in the algorithm for fitting the "spheres/targets" in the point cloud. Unfortunately, according to Mechelke et al. (2007), "the influence of these errors cannot be determined separately". Moreover, some laser scanning equipment can have issues through reflectance out of certain materials like marble and gilded façades.

It has been found in the examinations of the 3-D assessment in the field by Mechelke et al. (2007) that the range value (starting with the lowest to the highest deviation value), which is effected through the estimation accuracy of the instrument, as well as through the algorithm suitable for the sphere, shifted from

41 mm to 76 mm for the scanners of Trimble GX, Leica Scan Station, Faro LS 880HE, and IMAGER 5006.

In Mechelke et al.'s (2007) study, it can be observed that just the "time-of-flight" technique reached up to +6 mm within the derived distances.

The examinations of the accuracy, with reference to distance measurements in contrast with reference distances, showed that the outcomes provided the accuracy requirement of the manufacturer. However, this accuracy is somewhat unique for every instrument.

It can be noted that the laser scanning accuracy determination is different than the total station, which cannot depend on a single target measurement and can be difficult to reproduce. However, focusing on objects through known properties within the laser scan volume, for example reference spheres, can lead to utilisation of an appropriate component that relates directly to the output model and its purpose. Rather than evaluating each point, the model can be evaluated (Thomson, 2016).

Additionally, according to the guide of the 3-D Laser Scanning for Heritage by (Heritage, 2011), both levels of accuracy and resolution of measurement are typically related to object size and the purpose of the laser survey. Since there is no specific level required, the best guide to follow is "the best that you can do".

2.5.8 *Data integration*

In the case of Building Information Modelling, data integration refers to the process of connecting different formats of project information into one file format. This file format is suitable for working across multiple platforms. Moreover, it has been noticed that the interoperability between the same vendor's applications, such as Autodesk, can be done fluently. On the other hand, huge numbers of stakeholders' programs are needed to have a common platform for the design and for the analysis of the structures in different enterprises.

Many of the companies involved in the building industry very often work with program applications that are determined by their operability and the cost, which can cause many complications and difficulties in the case of transferring the data between different departments involved in the project's construction process (Darie, 2014).

A number of researchers have focused on the pre-data processing and integration of information in BIM for heritage buildings; however, exchanging and integrating the data faces a number of issues in cases of transferring digital-born material into the BIM environment (e.g. Green Spider plug-in for Autodesk Revit 2012, as shown in [Garagnani and Manferdini, 2013]). On the other hand, the issue of post-data processing for an attribute to be improved and integrated into different datasets remain unresolved (Saygi et al., 2013).

One of the proposed solutions for solving this issue is using a standard format, such as COBie standards, which is currently being applied in the UK. Moreover, the COBie standard (which is an abbreviation for the Construction Operations Building Information Exchange) is a spreadsheet data format that includes all the useful information regarding the building in a digital format (BIM Task Group, 2013).

Another standard is the IFC (Industry Foundation Classes), which was developed by the International Alliance for Interoperability, the main purpose of which is to cover both geometries and semantics (Kolbe et al., 2008).

In the last few years, a number of new advancements have been made in both commercial and open source software solutions for post-data processing and data management. Some of these solutions can be found in the Building Information Model server. Depending on the format of the IFC, Open BIM movement was introduced via Graphisoft, Tekla, and building-SMART, which aims to improve the quality of the building data that is linked with elements or complete buildings. On the other hand, improving the attribute management up to this point has not played a key role in the offered solutions.

Consequently, the external attribute data integration has not yet been fully discovered. The heritage architectural object requires an integrated representation within several thematic drawings. This can be described as the multi-layered modelling approach, which can include themes such as alterations, structural system, and material usage, and allows for them to be represented in a multi-layered 3-D environment. Even though Building Information Modelling offers temporal representations (4-D), some other thematic representations are beyond the abilities of today's Building Information Modelling approaches (Saygi et al., 2013).

2.5.8.1   HBIM and database

The main concern of the integration between heritage BIM and the database is to provide the ability to export the HBIM Revit model data to the database. Then, it is possible to allow for any change to the data and to import it back into the Revit model, which can be described as "bi-directional linking". This can allow for broadcasting this database information in a custom-coded (by employing software languages, for example, HTML, ASP.NET, Java, and VB), lightweight, online web platform that provides an interactive real-time model to be viewed, modified, and manipulated by the end users, "in this case UNESCO users", with regards to the heritage building (Pike, 2012). In addition, this will allow the involved organisations to streamline the update and delivery procedure.

The owners of the heritage building and the municipality users can have access to make reports, changes, and modifications within the web platform, and data changes can be directly filtered back up into the HBIM model. This integration can help with meeting the UNESCO and the World Heritage community's requirements, provide for a long-term maintenance plan, as well as aid with the organisation and control of the missions among the involved organisations.

Additionally, in the case of the Revit database, the information can be presented in a table view that can be edited before importing. Also, creating shared Revit parameters with the ability to add new fields can be offered by the table view. Moreover, every change made to these new fields within the database will directly update the Revit shared parameters over the future imports.

There are a number of ways that a Revit database link may be able to support the management of projects. For example, managing the prototype project is one of these ways. Through exporting the Revit prototype project data into the database of Microsoft Access, and Microsoft Excel, this step can allow for the modification and updating of the database through specific information, for example, the condition of the site, as well as the information relating to facilities management (FM) (Autodesk, 2016a, 2016b). Afterwards, this updated data can be imported back into the project via the Revit database link, which is subsequently distributed to the involved and responsible organisations. On the other hand, until recently, there were some technical issues regarding the connecting of the 3-D model to Microsoft Excel because of the clashes between the supported operating systems, Revit versions, and the applications' updated versions (Darie, 2014). According to Fregonese et al. (2015), "Revit is a great database that allows for collecting all information, planning or not, related to a 3-D object"; moreover, "Revit can be used to visualise the object development, but, it is much more difficult to use it as support for successive decisions" (Fregonese et al., 2015, p. 78).

An example of the integration between BIM and the database can be found in the BIM3DSG system presented by (Fassi et al., 2015). This system provides the opportunity to contain both the model, as well as the data, within a unique database outside the software. Furthermore, sharing the database can offer each (architect, civil, and electronic) sector in the project the ability to be involved in the stage of any decision needed and to follow the development of the project. This system of management for the project may possibly be useful and interesting for public authorities who are responsible for heritage management or to each person with specific responsibilities in the design or renovation sector.

Likewise, the BIM3DSG system can offer an easy way to generate and manage information within the database that has been developed. Indeed, personalising the information tables is dependent on the specific requirements of a public authority, as well as connecting the public to the model.

Most likely, the transition to such a method requires professional skills and knowledge of the modelling field; however, a little education in terms of specialised staff for phases can be required during the experimented process in order to create the model (Fregonese et al., 2015).

2.5.8.2   HBIM and GIS

Many recent projects have demonstrated the importance of using a 3-D Geographic Information System (3-D GIS) in the field of cultural heritage preservation (Dore and Murphy, 2012; Jayakody et al., 2013; Lorenzini, 2009; Saygi et al., 2013; Wua et al., 2013). A 3-D GIS is a system that allows for the managing, analysing, and visualising of 3-D spatial data (taking into consideration their topological relationships), in addition to the thematic properties of objects in the real world (Zlatanova, 1999).

The thematic properties of real-world objects can be easily and efficiently joined to the geometric representation of 3-D spatial data through relational databases. In addition, 3-D GIS systems are characterised by their "layered" architecture, where each layer contains a subset of features that represent the same theme (roads, water streams, buildings, doors, windows, roofs, etc.). According to Wua et al. (2013), 3-D GIS platforms are good alternatives for generating convenient and low-cost BIM and HBIM. In fact, these platforms allow for savings of storage space, efficient retrieval of data, and advanced spatial analysis. Another interesting project was conducted by Dore and Murphy (2012), where their idea came from integrating HBIM and 3-D GIS. The proposed method consisted of two main phases; the first one was to create a 3-D model by using the HBIM architecture library and the second one was to integrate this 3-D model into a 3-D GIS for more advanced analysis.

The concept of the integration between 3-D BIM and the 3-D Geographic Information System (GIS) provides semantically rich models, and offers benefits from both systems to help with documenting and analysing cultural heritage sites. In fact, BIMs (and HBIM) offer powerful tools to automate the modelling process, thanks to their parametric objects. However, they lack the capacity to add new attributes that are non-architectural (Saygi et al., 2013). In contrast, 3-D GIS has a great potential to easily integrate new information, in addition to analysing and querying spatial and attribute data. Also, dealing with a huge volume of historical building information requires high capacity and flexibility in updating the database (with reports, images, etc.), which can be an issue in the BIM environment. However, the major drawback of 3-D GIS platforms is their limited functionalities regarding 3-D editing (Saygi et al., 2013). 3-D GIS can create and deal perfectly with simple 3-D geometric elements; however, creating complex 3-D models with 3-D GIS is still an issue. However, this issue can be solved through the integration of HBIM and 3-D GIS.

In the case of heritage buildings, the HBIM has to be semantically enriched in order to take into consideration some important aspects that may be non-architectural. For example, it is necessary to offer information about the materials used in the composition of the building's parts, the historical context of the building, and the history of reparations that took place in the building.

2.6   Summary

The second part of this chapter has highlighted the critical importance of OUV that is an issue affecting nomination files submitted to UNESCO, with reference to inclusion in the list of World Heritage Sites (WHSL). Therefore, all sites or properties submitted for this process are subject to detailed examination of their potential for OUV that relates to their intrinsic value, and how sites are planned to be managed, sustained, monitored, conserved, promoted, and protected.

The third part gave an example of the UNESCO World Heritage nomination file, which related to Historic Jeddah, Saudi Arabia. Historic Jeddah has unique architectural characteristics, such as the Roshan, as well as the construction methods, which were common in the Red Sea area between the 16th and 19th centuries. Over time, the historic city has lost its brightness and has been neglected due to many reasons. As a result, many historical buildings are at risk of collapsing and eroding due to ageing and human factors, as well as disasters such as fires and floods. Many of these buildings are already abandoned and have become hideouts for criminals.

During the last decade, the local authority has paid a huge amount of attention to solving the situation in Historic Jeddah. In achieving this aim, including the Historic City in UNESCO's World Heritage List was the best solution for gaining worldwide attention for the historic city. Through the process of nominating Historic Jeddah and the Gate to Makkah to the World Heritage List, the SCTA and Municipality of Jeddah expected that the generated positive dynamics would contribute to the revitalisation and preservation of the historic district. In 2013, UNESCO approved the nomination file of Historic Jeddah but with a lot of conditions and requirements.

The fourth part gave an introduction to Building Information Modelling (BIM). BIM is perfectly suitable for providing the kind of data and information that can be used to improve the new design and building performance. On the other hand, applying BIM in the heritage field is still a new area for investigation (Wong and Fan, 2013). The concept of Heritage Building Information Modelling (HBIM) was introduced in the fifth part. In this part, more comprehensive descriptions were given concerning the approaches of other researchers like DIT, CIMS, ABC Department, 3DOM-FBK Trento, and CNRS.

The next chapter will argue for the use of BIM in the heritage sector, as a way in which to provide a new method to document and manage historical sites, and also discuss how Heritage BIM can meet the requirements of UNESCO's World Heritage nomination file.

3 The new model for UNESCO's WHNF

3.1 Introduction

As UNESCO's World Heritage Resource Manual states, "there are many different ways to prepare a nomination and it is not appropriate to offer (recipes) or to recommend a preferred working method". However, "the advice provided" in the manual "is intended to offer basic principles and guidance to assist States' Parties as they establish a working method". This could provide validity in order to improve and create other methods for preparing the nomination file.

Besides, according to Pannell (2006), "In creating World Heritage, the convention also creates, in many ways, a world without borders", and further, "As a highly organised global response to the myriad of local challenges facing the world's heritage, the Convention forms part of the new architecture of global governance". Additionally, as such, with many systems in the world that have achieved significant recognition, and UNESCO's World Heritage nomination system being one of them, they are facing several challenges and issues, and are continually adjusting their policies and strategies in order to address those challenges.

Rao (2010, p. 165) states that, "Clearly, as a part of the reflection process, it is time to review and rethink some of the strategies and procedures that have evolved over the years for implementing the Convention", and additionally, "Most of these procedures have become so complex that they require special efforts for building the capacity of persons in the States' Parties who are charged with the implementation of the Convention".

There are many issues and challenges in preparing and examining UNESCO's World Heritage nomination files, such as in the case of the huge numbers of heritage sites around the world, thus it takes a long time to prepare the sites and prepare the file, as well as sharing the data and communication between those involved in such heritage projects (Bolla et al., 2005; Rao, 2010; Szabó, 2005) .

This chapter will discuss the issues related to preparing UNESCO's World Heritage nomination file (WHNF), and why it is important to have new methods for preparing the nomination file, and how Heritage BIM can solve the issues in providing all the required information regarding these sites that will be listed in the UNESCO World Heritage Sites List (WHSL).

This will be done by examining the following questions:

1   *How can Heritage Building Information Modelling (HBIM) provide innovation in creating information missing for UNESCO's World Heritage status? And what additional cultural value can a sustainable update of HBIM provide for such sites?*

In response to this main question, the following will be discussed: what are the pieces of **missing information**, and what are the **issues with the WHNF** (see section 3.2), and what are the **current solutions** for solving this matter (see section 3.3)?

a   For which aspects of the UNESCO nomination file can HBIM provide highly accurate fully documented information at the scale required for the UNESCO nomination project?
In response to this question, the following will be discussed: what is being **done already** through "using HBIM" (see section 3.5), and what are the **opportunities for HBIM** (see section 3.6)?

b   How can HBIM be used to manage and monitor historical buildings?
In response to this question, the **practical use/applications of HBIM** will be discussed in chapter 4.

c   How can HBIM be used to better maintain, protect, and record the updated information of the historical buildings?
In response to this question, the design of the **JHBIM model** "case study" will be presented.

d   How can HBIM improve the cultural value of heritage buildings in the short, medium, and long term, as well as provide a better future for historical buildings?
In response to this question, **the lifecycle of the historic site** within HBIM will be discussed (see section 3.8).

e   How does HBIM impact on Digital Heritage?
In response to this question, the **future of HBIM and Digital Heritage,** and how they can **complement each other,** will be discussed (see section 3.9).

Part five of this chapter will discuss HBIM as a new method for documenting and managing historical sites, as well as the benefits of HBIM for the historical field, the challenges for HBIM, and the level of historic detail modelling. The approach of remote data acquisition through the use of TLS and the following processing of the survey data will also be presented. Lastly, the chapter will investigate the use of HBIM as a model to meet the requirements of UNESCO's WHNF. The HBIM method has been adopted for a case study of Historic Jeddah, which is described as Jeddah Heritage BIM (JHBIM) that focuses on Nasif Historical House, considered to be one of the most important historical houses in Jeddah. The main reason for choosing this case study is due to the huge gap in knowledge in relation to heritage buildings in Jeddah. Many of these buildings have no engineering data in order to enable them to be rebuilt in the case of collapse or any disasters.

3.2 The missing information and the issues with the WHNF requirements

During the last 50 years, since the UNESCO Constitution was established in 1945, and since the Global Convention Concerning the Protection of the World Cultural and Natural Heritage was established in 1972, many challenges and issues have emerged in relation to providing the UNESCO World Heritage nomination files (WHNF) in different ways. These issues include environmental, social, economic, protection, and management, together with the long time it takes to prepare such a file.

Many experts had discussed the issues and some have tried to solve them, such as Bandarin (2007), Rao (2010), Riccio (2014), Santana-Quintero and Van Balen (2009), and Szabó (2005). Even a country such as Italy, which has the largest number of World Heritage Sites (WHS) with 51 sites in total, and almost with no site in a dangerous condition and with all of their experience, faces numerous issues in terms of protection and management and in fact, recognises that these are complicated heritage sites (Riccio, 2014).

Through analysing the issues which are discussed by Bandarin (2007), Rao (2010), Riccio (2014), Santana-Quintero and Van Balen (2009), and Szabó (2005) of WHNF, the challenges can be divided into four main topics. The first topic regards the heritage site itself. The second topic is with regards to the people involved. The third topic relates to the UNESCO procedure. Lastly, the fourth topic is related to technical aspects, time, and cost issues.

3.2.1 Heritage sites issues

The most important issue that faces the nomination file is if there is any missing information regarding the heritage site. UNESCO requires certain information to be provided so that the reconstruction of the sites is possible in the case of any disasters. When accurate information regarding these sites is provided, these World Heritage Sites can be protected, and future generations can enjoy them. The lack of such essential information often leads to serious consequences, such as the refusal of the nomination file. Unfortunately, most of the heritage sites do not have any documentation made during the time they were built.

Many of the documents that are provided in the nomination file were made decades or centuries after the site was constructed. This issue can cause a lot of misinformation in terms of the heritage site. For example, in the case of Historic Jeddah, there is a big gap in prior knowledge in relation to the Hijazi heritage buildings. Many of these buildings have no engineering data for them to be restored in the case of collapse or any other disasters (Figure 3.1 and Figure 3.2).

Figure 3.1 An example of the issue of damage on-site for Historic Jeddah, Saudi Arabia
Source: Author

According to Engineer Sami Nawwar (2013), the Director of Jeddah Historic Preservation Department, Municipality of Jeddah, this gap was caused by two main aspects. The first aspect was the lack of attention to the economic and cultural value of these buildings by the locals. The second aspect was that there are not many experts in the field of Hijazi buildings.

Figure 3.2 Example of the challenges faced the fieldwork in Historic Jeddah
Source: Author

In many cases, the misinformation and the shortage of knowledge about a historic site are the main reasons for the building's destruction (Santana-Quintero and Van Balen, 2009).

More importantly, over the last decade, the common method for the documentation step was the traditional survey method through providing some images of the site and some sketches, which can result in a lack of credibility. Regarding preparing a UNESCO's World Heritage nomination file, the most common issues can be the complexity, management, and ownership of these heritage sites. Moreover, managing these heritage sites, including the ownership, are very problematic issues (Szabó, 2005).

The next most common issue is how the heritage site is being prepared to be investigated. Engineer Mohammad Yosof Al-Aidaroos, the Supervisor of Archaeological and Historical Sites at the Supreme Commission for Tourism (Saudi Arabia), claimed that during the preparation of the nomination file of Historic Jeddah, many locations in the historic city were not in good enough condition for fieldwork or for any investigations to take place (Figure 3.1 and Figure 3.2).

Minor issues can include the location of these heritage sites, as some of them are in hotspot areas and are a target of military action, such as in some areas in the Middle East and Africa. This can cause a lot of difficulty in terms of preparation or even examination (Bandarin, 2007).

3.2.2   Issues with the people involved

This includes everyone involved in the World Heritage nomination process, starting with the local residents to the decision-makers. Many experts such as Engineer Mohammad Yosof Al-Aidaroos pointed out that during the preparation of the nomination file of Historic Jeddah, those involved had very little experience and the missions were not clear for the participants.

One of the main issues with Historic Jeddah's first nomination file was, as Ziyad Al-Dirais, UNESCO's Kingdom Representative stated, "The dismissal of the Kingdom's application was purely down to technical reasons, as Jeddah's historical sites had been subjected to negligence and misuse from people who did not recognise its value", besides, "the negligence had greatly influenced ICOMOS's decision". This can lead to the influencing of the communities' behaviour in different ways, for example, in which way they think about developing these sites, and according to Cor Dijkgraaf (2003), in many of the worldwide heritage sites, "local communities and authorities often do not perceive the conservation and renovation of these heritage sites as a priority".

Wherefore, Bandarin (2007, p. 234) suggested a number of ways in which to solve these matters, such as "raising the level of management and human skills for conservation; and finally, informing the public of the achievements and challenges ahead".

Many owners believe that these buildings are not useful as they cost a lot to maintain, have no material return, and are generally not supported by the government. Therefore, they prefer to abandon them until they completely deteriorate and collapse, so that they can build new towers or commercial structures that generate a better return. The owners of these buildings see this option as more attractive and rewarding than preservation due to the high land prices that can be invested in for much better returns.

3.2.3 UNESCO procedure issues

There is a huge gap in knowledge in terms of the running and understanding the tasks of UNESCO's World Heritage nomination file, and even in understanding some important steps, such as the outstanding universal value "OUV". Additionally, by focusing on the WHNF process, many difficulties and issues can be noted, which can include UNESCO's decisions themselves. As Kishore Rao (2010, p. 166) claimed, "the extant process is contrary to the real intent of the convention of identifying and conserving heritage of outstanding universal value through a system of international cooperation and therefore, it does not contribute effectively to realising a representative, balanced and credible World Heritage List".

The World Heritage Sites List contains a vast amount of sites, amounting to more than 1,000 properties (see Figure 3.3). Rao (2010, p. 166) stated that, "this issue is certainly of genuine concern" as it can reduce the weight of the World Heritage label. Further, he claimed that, "as a part of the reflection process, it is time to review and rethink some of the strategies and procedures that have evolved over the years for implementing the convention. Most of these procedures have become so complex that they require special efforts for building the capacity of persons in the States Parties who are charged with the implementation of the

Figure 3.3 The World Heritage List updated in October 2016

Convention" (Rao, 2010, p. 167). Another issue, as Dijkgraaf (2003) stated, is the "Methods of conservation, maintenance and management of sites in developing areas of the world also differ greatly from those employed in the developed world". Likewise, Szabó (2005) claimed that, "The World Heritage Management Plan has a peculiar status in the present situation. Unlike all other plans discussed so far, it has no enforcement power other than the possible removing of the site from the WH List if the Management Plan is disregarded. It will really be a list of things every party agrees should be achieved. Ironically, this may be an asset: it could serve as a starting point that might, in the future, lead to unified management".

3.2.4   Technical, cost, and time issues

Regarding both the technical and cost sides of preparing such nomination files, Santana-Quintero and Van Balen (2009) claimed that, "The world heritage nomination process can be very costly for heritage organisations lacking funding and capacity". They describe the issues of preparing a nomination file as providing "Identification and description of the property; justification for inscription; state of conservation and factors affecting the property; protection and management; monitoring; and documentation". Additionally, they claim that the "identification through rapid-assessment heritage places is among the most important actions towards the protection of properties". Furthermore, they tried to solve this issue by using an "approach for carrying out rapid-recording assessments, in order to accelerate the identification, definition, and dissemination of essential information about heritage places for the enlisting on UNESCO's World Heritage convention. This rapid-recording approach can be used in case of pre-inventory, salvage recording, and recording in the case of emergency" (section 3.3 will discuss this further).

Even after inscribing the site into the list of UNESCO's World Heritage through the current process, there are many issues leading to disappearance these sites as well as these heritage sites can cost a lot, especially in the developing areas. One of these issues is the annual review report relating to how long and how much it will cost, and how it will be added to the nomination file. For example, by focusing on one of the nomination files, such as the Tower of London, a huge amount of paperwork can be found, images, reports, and 2-D drawings since 1987, and the amount may increase in the future.

From a technical point of view, the issue of how the local authorities can update or report on any future information and share this information with the UNESCO World Heritage List after it is submitted to UNESCO may be a serious issue in the near future and even in the distant future. This issue relates to the huge numbers of World Heritage Sites, "over 1000 WHS"; as Kishore Rao (2010) mentioned previously. It seems that moving away from a huge amount of paperwork and 2-D drawings to more advanced methods and techniques is a necessary focus and one that can't be ignored.

Many of the experts claim that these issues have affected the time necessary for preparing the nomination files. For example, regarding the Historic Jeddah nomination file (the case study), Engineer Mohammad Yosof Al-Aidaroos claimed that these issues relating to the heritage sites and participants involved result in a long time frame and a delay in submitting the final file, and in some cases, even affected the quality of the content.

3.3 Previous solutions for UNESCO World Heritage Sites

Studies have been conducted regarding the topic of UNESCO World Heritage Sites in relation to the criteria, OUV, states' parties, cultural governance as an approach for the purpose of the social and cultural sciences, politics of heritage, and the decision-making process (Dumper and Larkin, 2012; Jokilehto, 2015; Meskell et al., 2015; Schmitt, 2009). However, very few studies have been conducted on the topic of technological and information solutions and on developing a machine method to satisfy these requirements. This part will focus on some of the studies that have taken place in terms of discussing and providing solutions for the UNESCO World Heritage Sites and the nomination files. These studies have generally focused on the inventory plan, surveying, geospatial technologies, urban analysis, and documentation issues in order to provide information regarding the UNESCO World Heritage Sites and nomination file requirements.

3.3.1 *Proposed solution for pre-inventory plan*

The most important step toward inclusion in the WHL is the inventory plan or the inventory list, and a number of researchers have tried to provide solutions for the inventory step, such as Gillot and Del (2011) and Santana-Quintero and Van Balen (2009). Employing a photographic method, which is known as architectural photogrammetry, was the first solution, proposed by Santana-Quintero and Van Balen (2009). They demonstrated a method of finding out "rapid-recording" evaluations, which led to the speeding up of the definition, identification, and dissemination of the basic materials of the historic sites to be included in UNESCO's World Heritage List. Figure 3.4 shows the elements to record to prepare a baseline.

Santana-Quintero and Van Balen's method can be employed for salvage recording, pre-inventory, and recording in emergency cases. The aim of their approach is to provide a tool for heritage sites in cases of rapid-assessment in a relatively short time frame.

The approach is centred around "baseline recording" that is aimed at preparing graphic information, which is required to support the World Heritage nomination files. This approach can be summarised in five steps. The first step is to have a clear understanding of the site. The second step is identifying the location and the boundaries of the heritage site (Figure 3.5). The third step is to take a sequence of overlapping photos of the site. The fourth step is to upload all the data collected

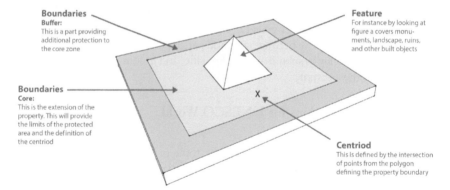

Boundaries
Buffer:
This is a part providing additional protection to the core zone

Feature
For instance by looking at figure a covers monuments, landscape, ruins, and other built objects

Boundaries
Core:
This is the extension of the property. This will provide the limits of the protected area and the definition of the centriod

Centriod
This is defined by the intersection of points from the polygon defining the property boundary

Figure 3.4 Elements to record to prepare a baseline
Source: Santana-Quintero and Van Balen (2009)

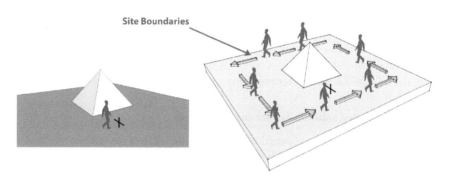

Site Boundaries

Figure 3.5 Property centroid and mapping the property boundaries

into a 3-D Earth Viewer and share this data with UNESCO and the stakeholders in order to obtain the approval of the heritage sites' boundaries. The final step is to build the 3-D models through using the photos and producing the required drawings to be used by UNESCO and the stakeholders.

Local authorities, governments, and international organisations can use the results of this method in order to capture sufficient and relevant information regarding the heritage sites. This method offers a preliminary record of the heritage sites with the possibility of sharing this information with others via the World Wide Web.

The benefits of such a method can be the enrichment of knowledge in terms of cultural heritage resources. Therefore, these World Heritage Sites can be protected, and future generations can enjoy them. Besides, as Santana-Quintero and Van Balen (2009, p. 3) claimed, if the Committee of UNESCO's World Heritage started "to investigate low cost approaches for preparing nominations, along with the current available technology; online free videos can be offered through social networks (such as YouTube educational) to provide a guide to organisations lacking resources to conduct assessments for potential world heritage properties".

3.3.2 *Proposed solution for inventory plan via GIS*

Illustrating GIS to provide an inventory plan and to make an analysis of the cultural sites was the second solution, which was provided by Laurence Gillot and Andre Del (2011) for the Oasis of Figuig, Morocco. The main purpose of the study was to offer inventory plans and to analyse the heritage properties of the oasis, in order to support the local government in preparing the nomination file of the oasis to be inscribed within UNESCO's World Heritage List.

GIS is known to be a powerful tool for management and analysis purposes that makes it achievable to generate thematic (historical, touristic, archaeological, etc.) inventories and mappings. Besides, GIS can be used as a shared tool with different applications, in the cases of heritage management, tourism management, scientific research, etc. (Gillot and Del, 2011).

When using GIS for archaeological purposes, Gillot and Del (2011) claimed that, "the application of the GIS in the archaeological domain turned out to be very useful in the programming of archaeological excavations, as well as in the interpretation of the data which arose from it". The data collection of this work commenced in three stages. The first stage involved the collection of planimetric maps and their georeferencing. Secondly, georeferencing was identified based on both GPS and Google Earth points. Lastly, the photo-interpretation of the Google Earth image was created.

The benefits of using GIS in this project allowed for the highlighting of endangered areas, particularly the ruins between 1917 and 2009, and many of the important archaeological discoveries in the area. Using GIS allowed the team to determine the boundaries of the Ksar's mosque (*Ksar* in Arabic means palace) and the boundaries the Ksar had in different ages.

Using GIS to manage this heritage site, moreover, offered the possibility of making an inventory of buildings and the important areas. In the case of using GIS for developmental purposes, a good example is in determining the most suitable location for a solar energy facility.

Through employing GIS for touristic purposes, the local authority could further focus on developing sustainable tourism to support the cultural development and socio-economic aspects of the site. Moreover, using GIS offered the possibility of developing tourist paths, which could include the most important places to be visited.

In this project, GIS potentialities have been demonstrated in different respects. GIS offers the possibility of collecting different scientific information and generating thematic mappings. Moreover, GIS demonstrates a huge capacity for supporting and using it as a management tool for heritage, scientific, tourist, and environmental purposes. As has been found in this project, to be listed on the World Heritage List, the use of GIS is very crucial in preparing the nomination file.

3.3.3   Proposed solution for dealing with the site complexity via DSM

One of the most common issues that can be faced in the historical sector is the complexity of historical sites. A number of projects have addressed this issue and attempted to solve it, such as Abdelkader El Garouani and Abdalla Alobeid (2013) in Faz, Morocco.

The main target of their work was "to generate a feasible platform for the detailed simulation of urban modelling". Digital Surface Model (DSM) was employed in this project, based on aerial image stereo pairs using the matching method in order to generate 3-D city planning. The project involved important phases, which included the restoration of the buildings, urban and landscape settings, all related to environmental and urban planning.

Moreover, GIS and the 3-D modelling software were employed as the platform to provide the virtual 3-D city model. This model can be used in several applications, such as city and architectural visualisation, navigation systems, urban planning, flood risk mapping, and urban risk modelling.

In the first step of this project, automatic image matching was used to generate the DSMs based on the imagery acquired from space and airborne sensors. El Garouani and Abdalla Alobeid (2013) employed the developed algorithms by Büyüksalih and Jacobsen (2006) and the developed procedure by Kwoh et al. (2004), in order to do this. The second step was to generate the 3-D city and urban modelling based on the DSM and the very high resolution (VHR) satellite images.

The 3-D models can offer a friendly interface for the users to query the urban environment such as the GIS, and to offer hyper-linked web-based information, and provide a model for visualisation and functional simulation.

Figure 3.6 shows the working process of DSM generation.

Several GIS applications can provide advantages from the use of generated 3-D city models, for instance, natural disaster simulation and telecommunications planning. GIS also offers a close examination of the repartition of interventions in the city, which assists understanding and guiding the mechanisms of restorations. Employing the GIS for Fez can be considered as a management tool for the purposes of the preservation and supervision of projected and implemented tasks.

3.3.4   Proposed usage of 3D-ICONS

3-D modelling has been commonly used in the last decade for different purposes, yet, using 3-D modelling for cultural heritage faces many challenges. One of the most important challenges is how these existing complex archaeological and architectural models can be adapted for the end users and for a wider audience, even if these heritage sites are 3-D real-life models based on artefacts, heritage sites, and monuments. The most important solution for dealing with this challenge was provided by D'Andrea et al. (2012) in the 3D-ICONS European project, which aimed to improve the Europeana content database for users by

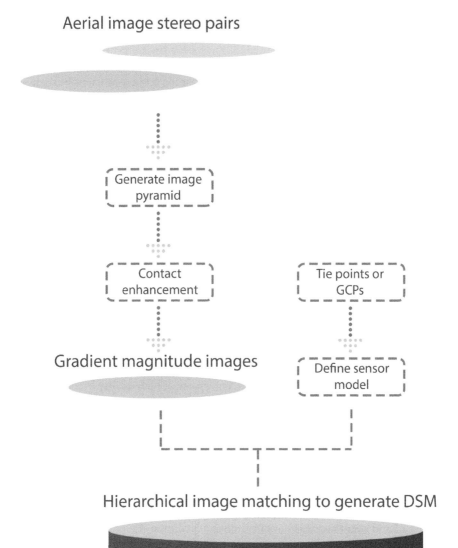

Aerial image stereo pairs

Generate image pyramid

Contact enhancement

Tie points or GCPs

Gradient magnitude images

Define sensor model

Hierarchical image matching to generate DSM

Figure 3.6 The working process of DSM generation

targeting European architectural and historical monuments to be made into 3-D digital models.

The aim of 3D-ICONS' project was to provide complete 3-D content, which was brought to Europeana through the project of CARARE that provided access for the public to the complex models, as well as increasing the masses' exposure to this attractive type of content (D'Andrea et al., 2012). The project began by surveying all the available data, the technical information, and the content gaps. During the next two to three years, the project of 3D-ICONS aimed to enrich the World Heritage Sites and the related objects for the purpose of visualisation through the Europeana project. As such, these projects can bridge the gap in the area of complex 3-D models, which are created for the purpose of scientific studies, making the 3-D content more exciting and attractive. In this project, three criteria were chosen for selecting the 3-D content in order to be employed within the 3D-ICONS.

The first criterion was cultural significance, the second criterion was how easy it was to transform the object into a user-friendly layout, for example, as a 3-D PDF, and the third criterion was the processing charge and resource usage. The 3-D models can offer many advantages for public users, such as visiting the models and the sites from remote locations, as well as offering a better understanding of the heritage site. For instance, by focusing on the model in a 3-D PDF, the model can deal perfectly with the small model size (under 20 MB) as the 3-D models can be contained by text, which offers content and descriptions for the model users.

The proposed method had five characteristics. The first characteristic was the flexibility of data creation. This flexibility could be determined via employing open source software and using different methods of 3-D data acquisition and modelling processes. The second characteristic was the advanced semantics, which was achieved through adding rich metadata to the digital content in order to meet the Europeana specifications. The third characteristic was the quality outcomes. During this project, the outcomes were of high quality from a cultural heritage perception. The fourth characteristic was business considerations for future implementation, which could include developing a cost analysis, IPR (intellectual property rights) management models, and organisational models. The fifth characteristic was attractiveness and wider interest, which could be determined by offering access to different heritage sites in Europe and the related information.

The expected challenges could be related to the most suitable format for representing the large, very complex 3-D models, and how the model could be reduced and simplified without affecting the quality or the potential enjoyment for the users of the model. Moreover, the software and the tools generally required in this project would be open source software and technologies. Additionally, the skills developed through the partners could allow them, personally or communally, to utilise expertise and consultancy assistances. This may have also required an update of technologies.

The expected results of the project were within three years on the time scale, including about 60 ICONIC sites and monuments being offered within Europeana, including hundreds of architectural details and buildings in 3-D models, and thousands of high-resolution images. Besides, 3D-ICONS would be qualified in order to offer a tried and tested procedure for the purposes of the production and conversion of the 3-D content and the related metadata. This metadata would contain the tools and technologies, guidelines for best practice, and wider knowledge built on tens of years of experts' experience. This would allow other European cultural heritage experts and developers of 3-D content to integrate their 3-D models within Europeana, to raise awareness and increase all the stakeholders' knowledge.

Unfortunately, this rich database of architectural details of the 3-D models created by Europeana was and still is not utilised by UNESCO. This is because they still haven't adapted the use of HBIM models in their system. The lack of usage of these models represents missing out on countless great opportunities.

3.4 The need for a new method for UNESCO's WHNF (the gap)

Since 1945, almost all of the previous solutions have focused on parts of the main issue but not tackling all aspects of it. As UNESCO's World Heritage Resource Manual states, "there are many different ways to prepare a nomination and it is not appropriate to offer (recipes) or to recommend a preferred working method". However, "the advice provided" in the manual "is intended to offer basic principles and guidance to assist States' Parties as they establish a working method". This could give validity in terms of improving and creating other methods for preparing the nomination file. Besides, according to Pannell (2006, p. 76), "In creating World Heritage, the convention also creates, in many ways, a world without borders", and "As a highly organised global response to the myriad of local challenges facing the world's heritage, the convention forms part of the new architecture of global governance".

As such, many systems in the world have achieved major recognition, and UNESCO's World Heritage nomination system is one of them. However, this system still faces several difficulties and issues and is continually changing its policies and strategies with the aim of tackling these difficulties.

Taking into consideration everything mentioned previously, the huge need for a new method or the integration of all the partial solutions is crucial. Thus, it is a necessity to come up with a common novel process in order to meet UNESCO's WHNF requirements and unite all efforts under one umbrella.

This part will discuss why is it worthwhile to find solutions for the issues at hand. The following points can give a clear picture of why it is important to find answers:

1 It takes a long time to get decisions from governments, local authorities, and the UNESCO World Heritage Committee. Moreover, lots of historic sites disappear without any recovery documents and even without any decision of the UNESCO WH to be in/out of the WHL (Santana-Quintero and Van Balen, 2009).
2 Lots of misunderstandings and misinformation during the nomination file preparation. This can be due to unclear missions and clashes between the participates' missions (Engineer Mohammad Yosof al-Aidaroos, the Supervisor

of Archaeological and Historical Sites at the Supreme Commission for Tourism, Saudi Arabia).

3   Providing a platform that provides the ability to follow up on the operation of the heritage buildings.

4   Supporting the legal systems of the heritage sites.

5   Supporting and providing the vision of a shared built heritage of the International Scientific Committees within ICOMOS. The main objective of this vision is to support the private/public organisations globally in cases of safeguarding, raising awareness, documentation, and management of shared built heritage (ICOMOS, 2013). Besides, supporting the integration of the heritage within the modern society in different ways such as in economic, social, and functional ways (Dijkgraaf, 2003). Providing an SBH vision can offer several benefits for the heritage sector (monuments, buildings, and sites) such as future maintenance, protection, conservation, rehabilitation, and enhancement (ICOMOS, 2013).

6   Allowing for more worldwide sharing experiences in the case of heritage management and protection systems, based on the recommendations, with reference to the conference "Linking Universal and Local Values: Managing a Sustainable Future for World Heritage" (22–24 May 2003, Amsterdam, the Netherlands).

7   Expanding the comprehension of the significance of human knowledge as capital, and as a basis for sustainable conservation and development, founded on respect and involvement for social and cultural values of local communities, and also based on the recommendations of the conference "Linking Universal and Local Values: Managing a Sustainable Future for World Heritage" (22–24 May 2003, Amsterdam, the Netherlands).

8   Providing a new method to unify the different sources in a single geographical reference system (Gillot and Del, 2011).

9   Providing a new method to deal with the new and the unique sources of information and knowledge for the cultural heritage (CH) sector. This can be provided to future generations (Santos et al., 2014).

10   Providing a new way to re-use the stored digital information related to cultural heritage. Furthermore, Europeana is one of the greatest examples of a large project that contains huge digital information about cultural heritage; however, the experience of reusing the information from this is minimal (Linaza et al., 2014).

11   Providing arguments that can influence the decision-making process with more transparency in decisions and with clarity. For instance, there is less concordance between the authority of the advisory bodies and the committee decisions, which can result in a lot of unclear decisions for the world heritage community (Luke and Kersel, 2012; Meskell et al., 2015).

12   Bridging the gap in the information management of heritage buildings and providing a collaboration solution between the information technologists and the conservation professionals (Saygi et al., 2013).

3.5 Heritage BIM, as a purpose for the WH nomination file

Now to answer the main question of this book: how can HBIM provide innovation in creating information missing for UNESCO's World Heritage status, and what additional value can a sustainable update of HBIM provide for such sites?

The concept of HBIM, data acquisition, and the integration of different data types will be introduced in relation to World Heritage. This will help to give a complete picture regarding the topic. Furthermore, with the information provided in chapter 2 and through analysing the UNESCO requirements, in cases of heritage buildings, it can be noted that the main keys to solving the issues and to meeting these requirements are via providing information in an interactive database. This allows a remote access and more comments and reports from different levels of users. Referencing sections 2.4 and 2.5, employing Building Information Modelling (BIM), can meet these requirements in an interactive 3-D application with the ability to integrate different types of data such as CAD, GIS, GPS, laser scanning, images, and much more data, which can be described as the Big Data and can be well structured, such as in the metadata.

3.6 The opportunities to meet UNESCO's requirements for heritage buildings within HBIM

Currently, many of UNESCO's requirements are provided in hundreds to thousands of printed papers and electronic documents in a PDF format. As mentioned earlier, there is not a standard form for providing the nomination file, which has led to a lot of misinformation, mistakes, and lack of clarity. This section will discuss the opportunities of HBIM within the boundary of UNESCO's World Heritage nomination file.

Employing BIM in the heritage sector, in order to provide fully documented information, has been done for a number of heritage sites around the world, such as in the projects of Barazzetti et al. (2015), S. Fai et al. (2011a, 2011b), Macher et al. (2014), Ma et al. (2015), Murphy (2012), Oreni (2013), Oreni et al. (2014), and Penttilä et al. (2007). Furthermore, by employing the HBIM for a heritage building, the lifecycle of the heritage building will be extended through being able to offer maintenance plans, protection and management plans, along with follow-up access (Fregonese et al., 2015) (section 3.8 will explain more about the heritage building lifecycle). Moreover, by focusing on the concept of BIM (in section 2.4) and HBIM (in section 3.5), the opportunities for HBIM to meet UNESCO's WHNF requirements can be found in the following ways, as detailed next.

3.6.1 *Providing the inventory plans*

The first opportunity of employing HBIM for UNESCO's WHNF is through providing inventory plans. Within the HBIM solution, these plans can be directly derived from the 3-D survey data from the laser scanning (Cheng and Jin, 2006).

In addition, more detailed structural data can be acquired via image survey data or photogrammetry (Santana-Quintero and Van Balen, 2009) and any accurate data that has been obtained before. This can produce a linear drawing and an orthoimage, which are suitable for preliminary inventory plans and for surveying heritage architecture that lacks structural records (e.g. outlines of the building plans and elevation). On the other hand, the HBIM model, which is based on the point cloud data from the laser scanning and the photogrammetry data, can provide more advanced inventory plans. For example, the HBIM model can provide the materials' quantities, and the mechanical, electrical, and plumbing "MEP" inventories.

3.6.2   Providing protection and management plans

The second opportunity of employing HBIM for UNESCO's WHNF is through providing protection and management plans. By having an accurate 3-D record on the heritage building's structures, such as 3-D laser scanning, that will ensure it can be used in the future for WHS (World Heritage Sites) status. This record can be used as a basis for the heritage BIM, as Henry Owen-John (2015), the Head of International Advice, Engagement Group, stated, it could be helpful, particularly when it comes to identifying how such structures can be protected and managed.

3.6.3   Providing the conservation documentation

The third opportunity of employing HBIM for UNESCO's WHNF is by providing conservation documentation. This can be automatically produced after completing the HBIM models. Moreover, conservation documentation can be defined as the visual and textual records that are captured through the process of the caring for and treating of the heritage building. These records can show the condition of the object, the history of the object's treatment, and any reports in relation to the object in the past or present. This can provide a complete story and as much information as possible about the object to be used in any future studies (Moore, 2001). Consequently, the conservation documentation of heritage buildings requires complete information and engineering drawings, which can be automatically provided via the BIM model, based on the laser survey data and the close-range photogrammetry. The output information from the laser and image survey data can be explained as an overturn engineering procedure, whereby a building's geometry, physical dimensions, and properties of the materials are captured in order to generate elevations, sections, orthographic plans, as well as the 3-D models (Cheng and Jin, 2006). In this case, the historic structures of the building are brought over from the design process but in the reverse direction, in order to illustrate the original information of the heritage building, which can include the original design, materials, and the structure (Murphy, 2012).

3.6.4   *Offering the standardisation form*

The fourth opportunity of employing HBIM for UNESCO's WHNF is through offering a standardisation form in order to prepare the nomination files. Through inputting the information into the BIM process, step by step, as is required, a common structure form containing the complete information that is needed can be generated. Providing a common standard for the nomination files can improve the efficiency, quality, performance, and reduce errors during the nomination process, as well as increasing the possibility of reducing costs.

3.6.5   *Providing building monitoring*

The fifth opportunity of employing HBIM for UNESCO's WHNF is by providing building monitoring in different stages of the building lifecycle. Through the BIM field management applications, heritage buildings can be monitored. As such, these applications have cloud-based collaboration, which can allow the on-site engineering staff to capture, report, and comment directly on the HBIM model in order to inform the decision-maker about the current situation. In this case, the HBIM model will be up-to-date and it also allows for any examination, analyses, or future studies.

3.6.6   *The integration of different data types*

The sixth opportunity of employing HBIM for UNESCO's WHNF is the ability to integrate different data types. This can meet a number of UNESCO's WHNF requirements in one model. As discussed earlier, a common standard format, such as IFC, can provide a wide range of information that can be exchanged easily within the BIM environment and with all the participants (Eastman et al., 2011).

3.6.7   *Providing maintenance and refurbishment plans*

The seventh opportunity of employing HBIM for UNESCO's WHNF is through providing maintenance and refurbishment plans for the historic building. This can be provided within the BIM facility management (FM) (Ilter and Ergen, 2015). As Del Giudice and Osello (2013, p. 226), claimed, "the need to refurbish the cultural heritage is becoming more important than the construction of new buildings". Moreover, refurbishment provides a great chance to improve the current situation of the heritage building, which can lead to a decrease in the operational cost of the building when conducted efficiently and effectively (Ahmad, 2014).

3.6.8   *Understanding the heritage building*

The eighth opportunity of employing HBIM for UNESCO's WHNF is by providing knowledge about the development and understanding of how the building has evolved and what is significant about it in historic and architectural terms. Moreover, it can highlight if the heritage building has significant artistic, historic, and scientific importance as a relic in the evolution of humankind. These heritage buildings need to be studied and understood in different areas such as in terms of building locations, construction and structural style, and the urgent need to protect these pieces of heritage (Cheng and Jin, 2006).

3.6.9   Sharing the information

The ninth opportunity of employing HBIM for UNESCO's WHNF is through providing access to the sharing and updating of data with UNESCO and stakeholders in an attempt to prepare the best and most suitable World Heritage nomination file. Besides, the integration of different sources of data and accessing them within the HBIM frame and application can offer new ways of updating the WHNF.

3.6.10   Providing the reconstruction of a 3-D building and object becomes reality

The tenth opportunity of employing HBIM for UNESCO's WHNF is by providing the reconstruction of a 3-D building and object for virtual reality. Through a comparison of the 2-D drawings, the 3-D HBIM model is characterised by abundant information and is more realistic (great visualisation). Consequently, modelling of this historical architecture can contain many important features, such as the construction style, as well as recording the shape and the structure. These 3-D models can provide important data for documentation and protection purposes.

3.6.11   Providing full engineering information

The eleventh opportunity of employing HBIM for UNESCO's WHNF is by providing full engineering information and drawings. Most heritage buildings have lost the blueprint copies or in some cases, the buildings were even built without them. Moreover, information and drawings, such as sections, plan details, and schedules, as well as 3-D models, can be automatically created within the BIM software (Murphy, 2012). This results in better visualisation of the heritage objects, which can be achieved via representing the 2-D and 3-D characteristics, sections, plans, elevations, and 3-D perspectives (Aouad et al., 2006; Eastman, 2006).

3.6.12   Providing the facility management (FM)

The twelfth opportunity of employing HBIM for UNESCO's WHNF is through providing facility management (FM) of the heritage building. This can be through supporting preventative conservation and providing the maintenance operations (Cheng et al., 2015).

3.7 The summary of HBIM to assist the UNESCO procedures

The summary of using Heritage BIM to assist the UNESCO World Heritage nomination file procedure is shown in Figure 3.7.

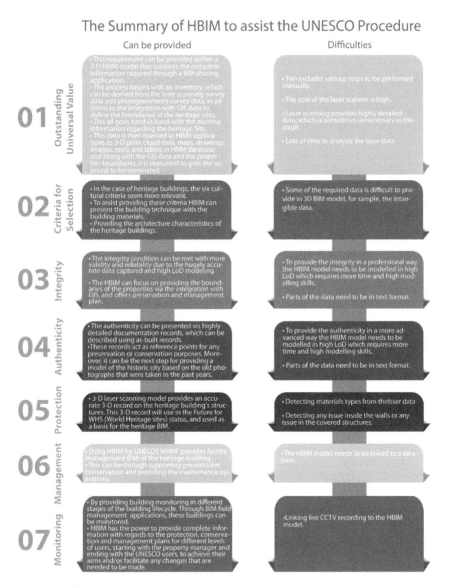

Figure 3.7 The summary of HBIM to assist the UNESCO procedures

3.8   Improving cultural value of heritage buildings

Determining the value of anything is always a reflection of the rewards it generates. By focusing on the heritage buildings, these rewards can be tangible, such as profit, and can be intangible, such as social benefits, for example, providing a great view and making the historic district look attractive. The heritage buildings that have tangible aspects are more likely to be conserved and preserved than the intangible ones (Navrud and Ready, 2002).

Generally, the heritage buildings around the world have very high value in themselves; however, sometimes the lack of preservation and conservation knowledge of these heritage sites can be one of the main causes of these heritage sites losing their value and as a result, they become unfit for human use. Consequently, they are abandoned and eventually collapse.

Employing BIM for the heritage buildings (HBIM) can improve the *value of heritage buildings in the short, medium, and long term, as well as provide a better future for the historical building through ensuring the lifecycle management of the building.*

Providing heritage buildings with a lifecycle approach can support different aspects during the timeline of these heritage sites. These aspects can include the facility operation and facility management (FM), which can support the construction management, project management, and cost management for any reconstruction or refurbishment for the heritage buildings. In addition, the lifecycle management can be used to examine the future development plans within the context of these heritage buildings. The HBIM model is created with all available data regarding the building within different periods "in the past, present", which can be described as a digital archive, and through the HBIM model, different operations can be undertaken in order to provide a better future for the heritage building (Stephen Fai et al., 2011b).

With the purpose of providing a lifecycle approach in regards to a heritage building within the BIM environment, a Heritage Building Lifecycle Management (HBLM) system is put into place. The HBLM puts into practice the level 3 BIM method, which provides a highly efficient extended collaboration model referring to the Heritage Building Lifecycle Management (HBLM) plan and the heritage manufacturing industry's best practice (Moriwaki, 2014).

Through providing the heritage buildings with a lifecycle and showing how these buildings can be managed in terms of different aspects such as conservation, operations, and rehabilitation of these heritage buildings for better uses, these plans result in the value of the heritage buildings increasing.

3.9   The impact of HBIM in Digital Architectural Heritage

During the last decade, technological developments have had a huge influence on many aspects of life, such as communication, transportation, and education. Additionally, this influence has reached new areas, such as the architecture industry and the heritage field. Since the Expedition Missions to different heritage sites worldwide have taken place, such as those to Giza and Luxor in Egypt or to Petra in Jordan, the architecture and survey techniques have been strongly utilised. For example, many of these sites had been surveyed and documented in recent times using the basic architecture and survey tools such as transit and theodolite, and total station. Through the enormous developments in architectural and survey technology, their importance been reflected in their expanding use, which includes their use in the heritage field. Over the last few years, the use of digital photos for close-range photogrammetry has been introduced in several historical sites around the world so as to capture the details about historical objects. 3-D laser scanning has been introduced to the heritage sector in order to scan and create accurate as-built 3-D point cloud models. In this case, according to Garagnani and Manferdini (2013), these 3-D models that are created by the 3-D laser scanning method are suitable for the purposes of periodic checks with regards to the structure's conservation, and most importantly, for documentation.

Recently, huge improvements in BIM applications have allowed for the integration and conversion of the laser scanning data within these modelling applications, such as Autodesk Revit and ArchiCAD. This step provides huge opportunities for employing BIM in the heritage field and further, to the existing building sector. For instance, between the 2000s and 2010, the research on this topic (of using BIM for existing buildings and in the heritage field) was very limited and had only focused on the countries that were already adopting BIM in their systems. On the other hand, after 2011 until the present day, there are a huge number of projects and studies that have been published in a relatively short time. The interesting point is that a number of these HBIM studies and projects have been undertaken in countries that do not currently support BIM in their systems.

By focusing on the main purposes of these HBIM studies and projects, it can be observed that the common purpose is to provide digital documentation for the heritage site, which can in turn, be used for different purposes at the current time or for future analysis. In addition, based on the main objective of Digital Cultural Heritage (DCH), which aims to digitalise heritage for the next generations, it can be concluded that HBIM is the natural evolution of Digital Heritage or it can potentially be one of the main sources of DH. Through the implementation of HBIM into the Digital Heritage field, HBIM can provide new opportunities for the DH field, for example, HBIM applications can provide new Virtual Reality (VR) systems, which can allow the public to interact with the 3-D environments of these heritage sites (Aurel Schnabel and Aydin, 2015). The ability of HBIM's attitude to store semantic, inter-related information can be successfully applied to the cultural heritage sites that can be easily managed (Garagnani and Manferdini, 2013), providing wide access to and sharing of information and databases, which will allow the experts of CH to archive open, reliable, reusable, digital backups by themselves (Mudgea et al., 2007). The most important opportunities can be found through the integration of the different data sources in a common model.

3.10   Summary

It can be agreed that, as with many systems in the world that have achieved significant recognition, UNESCO's WHNF system is facing several challenges and issues, and these systems are continually adjusting their policies and strategies in order to overcome those challenges and issues. Regarding the response to the main question (*How can Heritage BIM provide innovation in creating information missing for UNESCO's WH status?*), the WHNF challenges and missing information were discussed (see section 3.2), which can be related to the heritage site, the people involved, UNESCO's procedures, technical and economic issues. During the last few decades, several solutions have been presented to solve these matters and to meet some of UNESCO's WH requirements (see section 3.3). These solutions have focused on the criteria, OUV, states' parties, and cultural governance approaches for the purpose of social and cultural sciences, politics of heritage, in addition to the decision-making process; however, a few studies have been undertaken in the areas of technological aspects and information, and in order to develop an automated method to meet the requirements. Through focusing on those solutions, the strengths and weaknesses of each can be noted. In contrast, if these solutions are able to be combined in a common method, it could be more effective. Thus, it is very important to develop a new common approach in order to meet UNESCO's WHNF requirements.

Regarding the response to the first sub-question (*For which aspects of UNESCO's WHNF can HBIM provide highly accurate fully documented information at the scale required for the UNESCO nomination project?*), introducing the theory of applying HBIM to meet the WHNF's requirements was presented (see section 3.5), which is based on developing an interactive method that contains the heritage sites and links with different HBIMs for each heritage site in the area. HBIM is an interactive approach that enables a move from traditional 2-D constructive representation, as well as the 3-D content models, toward Heritage BIM, in order to assist with the preventative conservation, information sharing, and knowledge dissemination from the heritage site. Through employing HBIM into the process of WHNF, many opportunities can be found (see section 3.6) such as providing inventory, protection and management plans, offering standardisation forms, as well as providing full engineering information.

The HBIM method has been adopted in the case study of Historic Jeddah, which is described as JHBIM, which focuses on Nasif Historical House (see chapter 4) in order to respond to both the second and third sub-questions: (*How can HBIM be used to manage and monitor historical buildings?*) and (*How can HBIM be used to better maintain, protect, and record the updated information of the historical buildings?*). The main reason for choosing this case study is due to the huge gap in the data and studies regarding the heritage in Jeddah. As with many ancient cities, such as Aleppo, Jerusalem, Lisbon, Athens, and Damascus, many of these buildings have no engineering data to assist them in being rebuilt in case of collapse or any other disasters.

To respond to the fourth sub question (*How can HBIM improve the cultural value of heritage buildings in the short, medium, and long term, as well as provide a better future for historical buildings?*), it can be noted that HBIM can provide heritage buildings with a lifecycle approach to support different aspects during the timeline of these heritage sites. These aspects can include the facility operation and facility management (FM), which can support the construction management, project management, and cost management for any reconstruction or refurbishment for the heritage buildings. In addition, lifecycle management can be used to examine the future development plans within the context of these heritage buildings. The HBIM model is created with all available data relating to the building within different periods (in the past, present), which can be described as a digital archive, and through the HBIM model, different operations can be undertaken to provide a better future for the heritage building (see sectin 3.8).

To respond to the last sub-question (*How does HBIM impact on Digital Heritage?*), section 3.9 discussed the future of HBIM and Digital Heritage, and how they can complement one another. By focusing on the main purposes of these HBIM studies and projects, it can be seen that the common purpose is to provide digital documentation for the heritage site, which can in turn, be used for different purposes at the current time or for future analysis. In addition, based on the main objective of Digital Cultural Heritage (DCH) or Digital Heritage (DH), which aims to digitalise heritage for the next generations, it can be concluded that HBIM is the natural evolution of Digital Heritage or it can potentially be one of the main sources of DH. Through the implementation of HBIM into the Digital Heritage field, it can be anticipated that HBIM can provide new opportunities for the DH field, for example, HBIM applications can provide a new Virtual Reality (VR) system that can allow the public to interact with the 3-D environments of these heritage sites (Aurel Schnabel and Aydin, 2015). The ability of HBIM's attitude to store semantic, inter-related information can be successfully applied to the cultural heritage sites that can be easily managed (Garagnani and Manferdini, 2013), providing wide access to and sharing of information and databases, which will allow the experts of CH to archive open, reliable, reusable, digital backups by themselves (Mudgea et al., 2007). The most important opportunities can be found through the integration of the different data sources in a common model.

The following chapter will argue the use of JHBIM as a new model for UNESCO's WHNF, and the implementation of the process for capture to JHBIM, as well as detail how it can provide all the required information to help with achieving inclusion in the list of UNESCO's World Heritage Sites.

4 Jeddah Heritage BIM and the case study

4.1 Introduction

This chapter will introduce the case study and the issues that relate to the historic district in Jeddah, Saudi Arabia. Next, employing BIM for the historic buildings in order to solve these issues and to meet UNESCO's World Heritage nomination file requirements will be discussed. Also, in this chapter, the architectural elements of the Historic Hijazi Building will be presented. They are used for a more realistic representation and as a basis for modelling structures, particularly those behind the scenes or inside the walls. Next, the methodological work will be presented. This will include the data capturing and the data processing steps. This will be conducted according to the JHBIM and the HAOL modelling process. The last step will explain how the requirements of UNESCO's World Heritage nomination file will be met through the JHBIM.

4.2 Working toward JHBIM

4.2.1 JHBIM framework

As a case study for providing a new model for the UNESCO World Heritage nomination file, Historic Jeddah, Saudi Arabia's submission, was chosen as the case study for this book, which will be referred to as Jeddah Heritage Building Information Modelling (JHBIM), focusing on Nasif Historical House, which is considered one of the most important historical houses in Jeddah.

The main reason for choosing this case study is the huge gap in prior knowledge relating to heritage buildings in Historic Jeddah. Many of these buildings have no engineering data attached to them, which could be used for restoration and rehabilitation actions in the case of collapse or any disasters. In addition, these buildings will be subject to examinations in the next few years for UNESCO's WHNF annual inspections.

The first step for representing the historic buildings in Jeddah for conservation purposes and for them to be examined by UNESCO's WH community is the use of 3-D parametric models. Creating links between the historical building volumes and the different structural elements will allow for immediate and enhanced perceptions of these buildings. However, building a 3-D model in detail is an expensive and complex process that requires a high standard of capability and competence, along with different and numerous modelling software (Brumana, 1990; Brumana et al., 1990). In this case, more advanced technologies such as laser scanning is required to provide a quick and accurate data capture. Also, employing more advanced automated or semi-automated modelling approaches can provide high product quality, and more reasonable cost and delivery project time (PDT).

The JHBIM model will be based on on-site survey data, which includes the laser scanning survey and image survey, as well as the Architectural Hijazi pattern books and any existing useful data. Figure 4.1 shows the stages of JHBIM generation.

Indeed, the preparation of realistic and accurate 3-D content models of Historic Jeddah buildings is strictly linked to the requirements regarding the planned interventions; however, in order for it to be suitable for the professionals in the field, the 3-D model cannot just be a 3-D surface model (Nex and Rinaudo, 2008; Remondino and Niederoest, 2004). It also needs to take into account the geometric rules of construction, the materials, the thickness of the walls, the organisation of elements, and the different construction periods for each aspect of the historical building's structure (Oreni et al., 2012).

The study of Historic Jeddah's buildings and their constructive methods begins with the observation of the current building, moving slowly through the survey to the virtual reconstruction of Historic Jeddah's building structures. Afterwards, the Jeddah Historical Building Information will be put into the "JHBIM" model for the UNESCO nomination file and what JHBIM can offer for both UNESCO and the Jeddah Municipality will be presented. The case study of this book will examine different methods for HBIM, then look at the best approach that can be used to meet UNESCO's WHNF requirements.

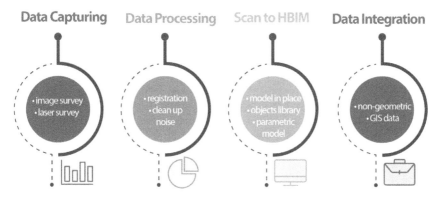

Figure 4.1 Stages of JHBIM generation

4.2.1.1   The first JHBIM step

This step will examine the on-site data collections methods based on the laser scanning survey and the image survey methods which have been employed in other studies (S. Fai et al., 2011a, 2011b; Murphy, 2012). This step will include collecting any existing information regarding the heritage site, such as any manuscripts, architectural drawings, reports, images, and historic information.

4.2.1.2   The second JHBIM step

This step will examine the "Scan to Jeddah Heritage BIM". During this step, the Autodesk Revit and Revit family will be utilised in order for a 3-D JHBIM model to be built. This will be done using the 3-D point cloud, which will differ from Murphy's (2012) method but is more similar to Stephen Fai et al.'s (2011a, 2011b) method. However, the main difference between JHBIM and Stephen Fai's case will be in the architectural style of the building. The prime modelling approach will be a manual technique needed for the difficult and non-geometrical shapes, while the semi-automated approach will be used for the simple and geometric shapes, such as for some of the walls and the floors. This step will begin after receiving the registered point cloud. The JHBIM will be linked to the data collected in the database, therefore each alteration to a parameter will change the shape of the elements as it is in progress (Boeykens, 2011; Boeykens et al., 2012).

4.2.1.3   The third JHBIM step

This step will develop a parametric objects library for the Hijazi objects, which will be described as the Hijazi Architectural Objects Library (HAOL). As such, in the research of Chevrier et al. (2010) and Murphy et al. (2013), such libraries' application improved the algorithms and methodologies for use in the data survey, particularly the point clouds, and for the model in the Building Information Modelling (BIM) software. As a result, it is necessary to focus on the level of detail and simplification of the models suitable for the conservation project. This is linked to the real opportunity for modifying the parameters of the architectural elements' shapes, thus exacting the historical objects that are often unique and irregular. The procedure of creating the BIM model will be different from Murphy's (2012) method, which was based on the Geometric Descriptive Language (GDL) in the ArchiCAD program. Overall, the HAOL will be based on the Revit family. The objects in the library will be highly detailed to a point where they are "as built level of details". The objects in this library can therefore be inserted into any future project in the historic area of Jeddah.

4.2.1.4   The fourth JHBIM step

This step will integrate the non-geometric data and any existing information such as restoration reports, GIS data, images and text data into the JHBIM model in order to satisfy the requirements of UNESCO's WHNF and to provide full engineering drawings and information. This engineering information can be used for all conservation and preservation purposes in the future.

The workflow of JHBIM can be found in Figure 4.2, with the process starting with the data collection step. This data will subsequently be evaluated by the administrator. *This step is similar to the justifying outstanding universal value (OUV) step.* Next, the surveying data, "TLS and image survey data", and any existing drawings will be used as a basis for the modelling, while the non-geometric data "such as the reports" will be input into the HBIM model gradually as needed.

Figure 4.3 focuses on the process seen in the dotted box in Figure 4.2, building the HBIM model and meeting the UNESCO WHNF requirements. During this process, an inactive 3-D modelling will be created in order to provide a model that contains all the required information. The main WHNF requirements will fall under four main sections. They are the inventory plan, the protection plan, the management plan, the monitoring, as well as non-geometric requirements, such as the integrity conditions' reports and authenticity conditions' report. These main requirements will have supporting data "such as the restoration plans, maps, satellite images, risk management and so on", which will be inserted and provided via the HBIM model.

Next, this model will be validated and tested and in the case of WHNF, it will be via *the IUCN* or *the ICOMOS.* If any issues or misinformation were found, they will be reported back to the HBIM model, and consequently the issues identified will be sent to the responsible department or person. In the case that all the information was found to be accurate, the model would be transferred to the *World Heritage Centre (WHC).* In this step, the WHC will use the HBIM model to demonstrate WHNF's requirements in a standard suitable form for the heritage community in order for the decision to be made with regards to the site. Figure 4.4 shows the integration between the workflow of JHBIM with the workflow of the UNESCO WHNF.

4.2.2   JHBIM level of detail and accuracy

As can be observed, the heritage buildings in Historic Jeddah have a unique architecture style. Also, these heritage buildings are different than modern buildings in several aspects. One of these aspects is the level of detail and the accuracy of the modelling. It can be argued that there is no official standard for the level of detail and the level of accuracy in the case of capturing heritage building in 3-D measurement techniques and in the case of producing this information in 3-D parametric models.

However, the requirement of accuracy that should be captured in the 3-D models can be determined by the way in which the heritage building and their architectural elements are understood. According to Letellier and Eppich (2015), the level of detail for the heritage conservation project in the case of 2-D drawings "can vary between approximately ± 2 mm and 5 mm for building elements and between ± 10 mm and 25 mm for building plans, elevations, and cross sections."

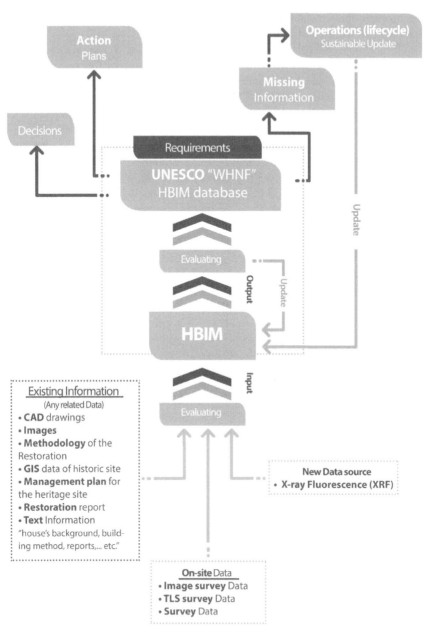

Figure 4.2 JHBIM water flow diagram

Source: Author

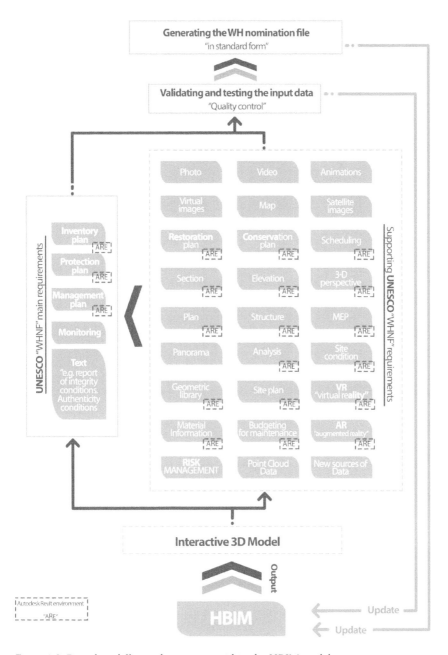

Figure 4.3 Providing different data sources within the HBIM model

Source: Author

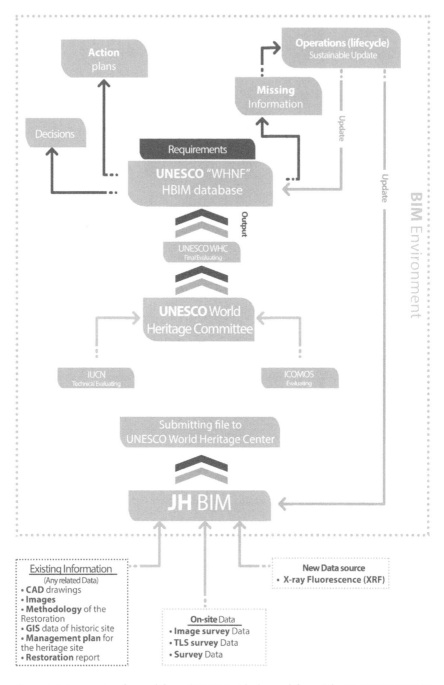

Figure 4.4 Integrating the workflow of JHBIM with the workflow of the UNESCO WHNF
Source: Author

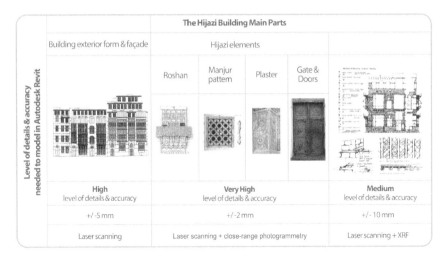

Figure 4.5 Level of details and accuracy needed to model in Autodesk Revit

Source: Author

This can be increased in the case of the (as-built) level up to (± 1 mm and ± 2 mm) for 2-D plans and elevations.

Through focusing on the heritage building in Jeddah, it can be noted that there are three main parts (i.e. building form and façade, Hejazi elements, and the building structure). Each of these parts needs a different level of detail and accuracy in order to be modelled in BIM form. Using the terrestrial laser scanning can offer a very high level of detail and accuracy. However, there is some limitations in scanning areas that are difficult to reach. For example, collapsed areas or over and behind the objects. In this case, it can be useful to employ the photogrammetry method to complement the task.

Figure 4.5 shows the Level of Details (LoD) and accuracy needed to model in Autodesk Revit for the JHBIM project.

4.3 The importance of the Hijazi architecture theme

One of the most important considerations regarding Hijazi historic buildings is the architectural style of these heritage sites. This is the main reason for the uniqueness of these buildings. This is due to a number of reasons. The first reason is based on the fact that they were built at a time when no developed mechanisms were available and with creative architectural ideas but with limited human capability. The second and most important reason is that this particular architecture style is common in different historic sites, especially in the Hijazi region and in the Red Sea area (see Figure 4.6). Moreover, this architectural style is evident even in different parts of the Islamic world (such as Historic Cairo, Baghdad,

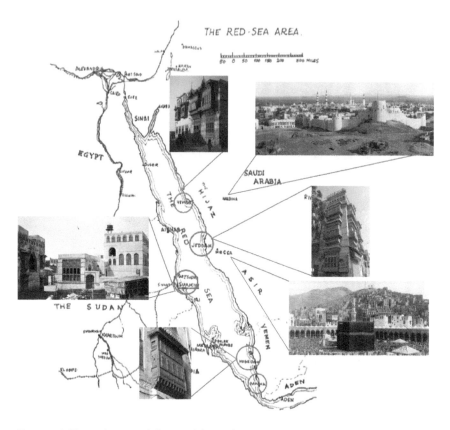

Figure 4.6 The architectural theme of the Red Sea area

Damascus), in Europe (Andalusia, Spain), and South America (Lima, Peru) (see Figure 4.7).

Unfortunately, there is a huge gap in engineering information regarding historic Hijazi buildings. Very few architectural books have dealt with Hijazi architecture. One example is *The Coral Buildings of Suakin, Islamic Architecture, Planning, Design and Domestic Arrangements in a Red Sea Port* by Jean-Pierre Greenlaw (1995). However, this book shows Suakin city in Sudan, which is on the opposite side of Jeddah, but on the African side. Suakin city can be described as the twin city of Historic Jeddah. Both Suakin city and Historic Jeddah have common architecture features between them. Figure 4.8 shows an example of an architecture feature that is common between the buildings in Historic Jeddah and Suakin. Another example is the book of Arab architecture, *Traditional Domestic Architecture of the Arab Region* by Friedrich Ragette (2003).

On the contrary, in the European modern age, many of these architectural rules have been agreed upon due to architectural manuscripts and the architectural

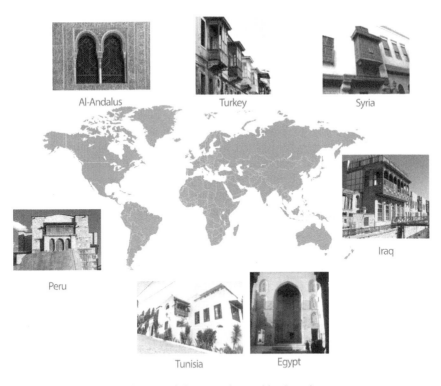

Figure 4.7 The Hijazi architectural theme in the worldwide scale

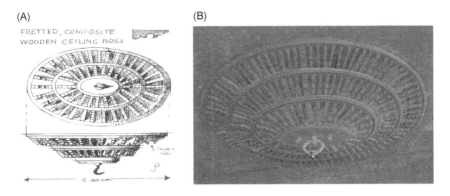

Figure 4.8 An example of the architecture features that are common between (A) Suakin historic buildings and (B) Jeddah historic buildings, "Nasif House"

pattern books. Indeed, these manuscripts and pattern books are considered one of the main resources for helping with understanding how these heritage sites were built, and a great example can be found in the European classical architecture pattern books such as *The Elements of Architecture* written by Sir Henry Wotton

(1968). These manuscripts and books are numerous in European and worldwide libraries.

The historic Hijazi buildings have a very unique architectural style and building structure. Therefore, in order to model such heritage sites, it is very important to understand the rules and order of the architecture elements.

By focusing on the Hijazi buildings, it can be noted that there are two main factors which can guide the design of the building. These main factors are the air circulation as well as the privacy. Almost 70% of the building façades are openings for circulation, in order to provide as much fresh air into the building as possible. Besides providing privacy, the Roshan, Mashrabiyah, and the wooden shish (Manjur) are present in these heritage structures. Throughout this chapter, the structural components of the Roshan, Mashrabiyah, and the wooden shish and the process of how they will be modelled will be presented. These architectural objects are found in the case study "Nasif Historical House" in Historic Jeddah.

4.3.1   Hijazi building structure and construction techniques

This degree of deterioration of Jeddah buildings offers valuable insight into the techniques used in their original construction, which are described by Greenlaw (1995) through the use of the examples of buildings of similar styles in Suakin city in Sudan, which is located on the opposite side of the Red Sea coast. This early written record of the buildings offers very important information regarding the building techniques, tools, lifestyles, and other types of unique buildings that have since disappeared. Coral has been used to construct buildings in different locations along the coast of the Red Sea; however, this coral varies significantly in type, size, and shape and was often rough cut to avoid reducing the size of coral blocks. However, when needed for door and arch surrounds, larger coral stones were cut for such purposes.

The common pattern of construction in Historic Jeddah and other such sites in East Africa, according to Orbasli (2007, p. 307), "was to use coral rag or irregular shaped stones that were then covered with a lime render".

It can be noted that the thickness of the walls of these building was greater at the base and then the thickness reduced gradually at each floor level, so while the common thickness was 0.8 m at the base, this became around 0.15 m less at each new floor level. Therefore, by the third-floor level, according to a local contractor for Historic Jeddah, Amm Saad (2011), "the thickness of the walls was around 0.5m; however, variations were likely as other findings suggest that the base thickness of coral walls of these buildings was around 0.6m".

Figure 4.9 shows that the method of building using the coral blocks, consolidated with timber, which is called "Takaleel" in Arabic, was done at approximately 1.2 m height that was laid horizontally. According to many experts, such as Professor Sami Angawi, the purpose of using "Takaleel" is to compensate for the coral blocks from expansion due to the changes in temperature, as well as improving the tensile performance of the walls. Furthermore, the timber was used to support the wall in upper opening areas such as windows and doors.

Figure 4.9 Example of Takaleel
Source: Author

According to Amm Mahfooz (2011), a local contractor at Old Jeddah, "Five-meters spans were achieved in Jeddah houses with uncut rough timber of around 0.2m diameter". When tall buildings were constructed, the weight of the structure was lessened by including semi-closed terraces and courtyards, and timber boards or canes laid diagonally over the beams, making up the construction of the floors and forming a diamond pattern.

On top of this layer was "Hasr", or palm leaf matting, and then a layer of coral limestone hard-core coated with lime for bonding, and finally a screed layer was applied. Tiles were also laid on floors in more recent historical houses (Amm Saad, 2011).

Several things can cause the deterioration of the coral walls, such as humidity, rain, sand storms, and salt. Therefore, plaster and lime renders were used to protect the coral walls, and this plaster would often be decorated.

4.4 Developing Hijazi Architectural Objects Library (HAOL)

In Jeddah and as with many historical buildings in Europe and around the world, building façades are of special interest. Moreover, the details of such windows, balconies, stonework, and ornaments give each historic building its individual character (Böhm et al., 2007). Therefore, each object of these historical buildings must be classified in an architectural objects library. Recently, a number of studies have been focusing on this topic in Europe and Canada, such as Fai et al.

(2013), Murphy (2012), and Oreni et al. (2013). The following parts describe in more detail the approaches of these object libraries.

4.4.1   Case studies of architectural objects library worldwide

4.4.1.1   The method of DIT, Dublin Institute of Technology (Maurice Murphy et al.)

Modelling the objects was based on using the Geometric Description Language (GDL), which is a language for creating parametric objects in the ArchiCAD BIM program. The scripting in GDL allows for sharing and editing of the parametric objects at different levels. The objects library functions can be used as a plug-in for BIM, and it is a novel prototype library of parametric objects built from historic data and a system for mapping the parametric objects onto a point cloud and image survey data. Moreover, the use of historic data introduces the opportunity to develop details behind the object's surface concerning its methods of construction and material makeup. Using GDL the classical elements detailed in the pattern books are reproduced using a design framework based on parametric rules and shape rules (Murphy, 2012).

The objects are scripted with parameters making them dynamic objects that can be reused. A bottom-up approach is adopted which starts with the smallest building objects such as ornamental mouldings and profiles. These uniform objects are created from a shape vocabulary of 2-D shapes that allow for all configurations of the classical orders. The shape rules are used to transform these 2-D shapes, representing classical mouldings and profiles. Non-uniform and organic shapes are developed in GDL through a series of procedures attempting to maximise parametric content of the objects. These shapes are stored as individual parametric objects or combined to make larger objects such as columns, pediments, walls, windows, or roofs (Dore and Murphy, 2013).

4.4.1.2   The objects library of CIMS, Carleton Immersive Media Studio (Stephen Fai et al.)

The objects library was built via the Autodesk Revit family based on the laser scanning point cloud. Early in the objects library modelling process attempts were made to reduce the amount of modelling and the number of models necessary to create the objects and establishing reference planes and reference lines within the Revit family modeller. The parameters were established to define how far the centre of the object frame is from the wall façade, as well as the height and width of this object. Although the same object types appear in several instances across the building, their dimensions from instance to instance can vary slightly in section and elevation. The single parametric version allows all size variations present from instance to instance to be compensated for by simply adjusting the given parameters of the model (S. Fai et al., 2011b).

Further benefits are seen through the ability to quickly upgrade all instances of that same object type at once while still maintaining all of the different size variations present across the multiple instances it is placed. Reference planes and reference lines within the Revit families act as a user-friendly means to adjust the geometry of the object manually within the Revit Family Editor by clicking and dragging the reference lines (Fai et al., 2013).

The number of parameters in the model have been reduced to the most fundamental parameters deemed appropriate for accurately recording the scale of each object type within centimetres of the accuracy of the TLS data. The high-resolution point cloud from the TLS data proved an effective template for placing and adjusting the scale of each object from instance to instance. The parametric capacity of BIM tools allows for an easier, faster, more flexible, and a more accurate modelling workflow. This procedure has also proven effective, as there is a consistent parameter naming format that is carried throughout all of the models. By knowing how to control one model, a modeller is easily able to modify all models using a consistent and familiar method (Fai and Rafeiro, 2014).

4.4.1.3 The objects library of ABC Department, Polytechnic of Milan (Daniela Oreni et al.)

The ABC Department creates the objects library using a number of methods and programs. Two different Building Information Modelling program platforms have been evaluated, Graphisoft ArchiCAD and Autodesk Revit, regarding the possibility to create irregular geometric objects that constitute covering structural elements. The Bentley's Point Tools plug-in for point cloud modelling, within Rhinoceros modeller and Leica tool for Revit, have been examined in order to create the object families starting from the LIDAR data (Oreni et al., 2013). The interoperability through Graphisoft ArchiCAD is on course of assessment, in order to express each room and zone, then to reconstruct the spatial aggregation between the object's components. Through comparing different BIM programs, the ABC group concluded that the ArchiCAD offers greater flexibility and adaptability in the modelling of irregular geometric objects, avoiding operations of model simplification too far from the object's real shape. Space Boundaries relational object will allow the single room unit to be related one to the other inside the whole building construction. The assessment of interoperability between the IFC output format file is on course of developing a program devoted to the thermal assessment and evaluation based on the object room component analysis. The export examinations of the geometry were intended to confirm the complete interoperability between the different programs using gbXML or IFC format file (Oreni et al., 2014).

4.4.2 The method of creating the JHBIM objects library

These Hijazi elements, such as Roshan and Mashrabiyah, have become the vocabulary of the Old Jeddah buildings. Since there is a huge gap in the Hijazi architectural library deeming it unable to provide these unique elements, the

Hijazi architectural objects library has been created for the JHBIM project. The objects library of JHBIM is connected to the data that was collected from the database, where each single modification of a parameter can lead to a change in the shape of the object. As a result, considering the level of detail is very important. It is also important that the object models can be simplified and easily modified in order to be suitable for the preservation plan, and for the models to have a greater opportunity to be utilised. This is particularly required for the buildings in Historic Jeddah, as the building objects are always unrivalled and irregular. According to Dore and Murphy (2013, p. 369), "Due to the individuation of the form, grammar and stylistic rules can be used to create a library of historical elements" and similarly, for Jeddah historical buildings.

The result will lead to both the building of an abacus of local construction objects and to match the objects' real dimensions with the information derived from any previous architectural drawings. This will make the models detailed in accordance with the real conditions. The purpose of this library is to use it as a plug-in for existing BIM software platforms such as Autodesk Revit. This plug-in can be accessed through the Add-Ins software ribbon, which is found in Autodesk Revit. Moreover, as such this plug-in can be developed through using Microsoft's Visual Studio. Through this plug-in the Hijazi Architectural Objects Library (HAOL) can be introduced, documented, and will support any future projects in Historic Jeddah.

The theory behind the HAOL originates from modelling these elements in 3-D modelling software, such as Autodesk Revit. The framework for creating the Hijazi Architectural Objects Library began with analysing and understanding the architectural rules of these elements. Secondly, understanding the purposes and the level of detail were a key in achieving the most suitable quality required for the model. Thirdly, the Hijazi architectural objects were classified in relation to a number of criteria such as the shape, the amount of detail, and the style of these details. In the beginning, the library was divided into three main types, "i.e. Roshan, Gate, and Manjur patterns" (Figure 4.10). Next, these important Hijazi object types were modelled. Through understanding and simplifying these objects and dividing them into main parts and sub-parts, this step allowed for a reduction in the complexity involved.

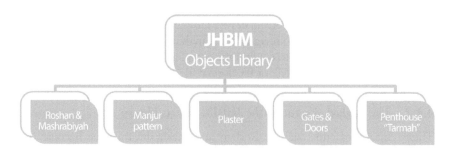

Figure 4.10 JHBIM objects library layout

The next step was creating the object's structural form and after that, modelling each single component as close to its real condition as possible, with "as-built level of details". The final step was integrating the multiple data in order to complete the final 3-D model of the Hijazi objects, which was appropriate for exportation to the JHBIM system.

Figure 4.11 shows the outline of the steps of creating the JHBIM objects library.

There are several programs that can be used to build the models of the objects library. In this stage, Autodesk Revit (2014, 2015, 2016, 2017 versions) was used. The Autodesk Revit software has a lot of advantages, such as rapid building and changes to the 3-D model, high-quality construction documents, and a high level of flexibility (Baik et al., 2013). In fact, to deal with such complex Hijazi objects in Autodesk Revit, the Revit Families were convenient. These files are stored in ". RFA" format and are able to be inserted directly into the Revit project. These RFA files are usually described as data files that hold one or more 3-D models. These 3-D models can be inserted into a 3-D scene and were generated and saved with the Revit Family Editor (ReviverSoft, 2013).

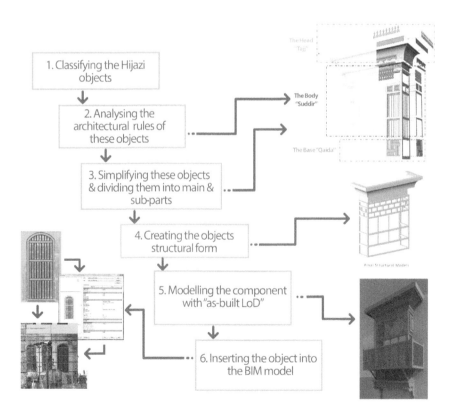

Figure 4.11 The steps of creating the HAOL

There are two methods for creating Revit family objects. The first method is by using the "Traditional Family Editor". In this method, the object must be sketched onto a 2-D work plan. The second method is by using the "Massing Family". The common stage among the two methods is saving the file in the ". RFA" extension; however, the main difference between both methods is that the massing environment is a 3-D work environment, which allows one to work directly in a 3-D view. However, according to Paul, "the mass category is only useful as a 'clay model' or 'study model' in the project" (Paul, 2013).

4.4.2.1 *JHBIM library shape rules*

By analysing and understanding the architectural rules of the Hijazi objects, and according to the Hijazi and Islamic architectural pattern books, it can be noted that the Hijazi architectural shape rules are always composed of basic geometric shapes such as boxes shape (cube and cuboid). These boxes are shaped and decorated to be suitable for the purpose for which they were created. For example, by focusing on the Roshan it can be noted that there are three main boxes (see Figure 4.12). These boxes are designed with ornament and geometric patterns. Moreover, the beauty of these Hijazi objects is based on the details of these decorations.

The Hijazi objects always have the same structure form, however, the difference is the decorations details and how they are assembled, which change from one building to another. An example is the Manjur pattern, made of specially cut laths of wood that fit into one another at right angles (criss-crossed). These laths of wood can be described in a basic geometric definition as a cuboid (see Figure 4.13).

The following parts will discuss in more detail the modelling process of creating the objects of the Hijazi Architecture Objects Library (HAOL).

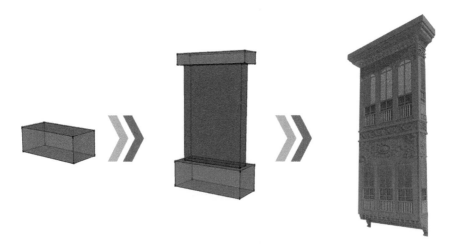

Figure 4.12 Evolution of the creation of the basic geometric shape of the Roshan

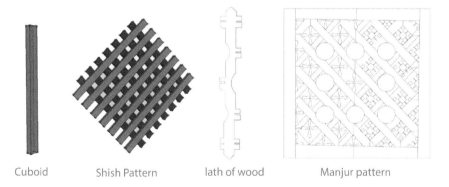

Cuboid   Shish Pattern   lath of wood   Manjur pattern

Figure 4.13 Evolution of the creation of the basic geometric shape of the Manjur pattern

4.4.2.2   *Modelling Roshan and Mashrabiyah*

The Roshan and Mashrabiyah are the most important characteristics in the Hijazi buildings. The Roshan and Mashrabiyah can be described as a "large project-ing wooden structure" on the Hijazi houses' façades, with a recognisable latticed component. Therefore, it is necessary to give more attention to them. Roshans and Mashrabiyahs are characteristic symbols of the architectural heritage of Jed-dah city: they are located almost all over the historic district's buildings. Roshans and Mashrabiyahs cast light and shadow on the houses' façades. These unique elements can always be found on the ground floors. The unit of the Roshan is created by a front and two side boards, and by a ceiling and a floor. The Roshan is created with wood, with mixed moving and fixed parts. The whole Roshan's structure is firmly tied into the brick by a series of wall brackets or consoles. These consoles are made so as to incorporate elaborate and, in some cases, polychrome ornamentation.

These wooden surfaces are carved and decorated. The parts are laid on top of console underpins, which are inside the brick. The side boards are "stapled" over the opening, utilising metal spikes that are attached to the wooden parts and tied into the wall.

In the lower section, the boards are built in grooves, constituting a perfect plinth. Overhead, from the interior, it can be observed that the eye-level is fitted with sliding wooden shutters, full or louvered. These sliding parts slide into the grooved base using a technique known as the guillotine technique.

On the higher level, Mashrabiyah parts are fitted to provide air circulation for the room. These structures are surmounted through a wide exquisite cornice.

By providing the as-built level of detail for both the Roshan and Mashrabiyah, it is possible to protect these Roshans and Mashrabiyahs from any further deterio-ration. Moreover, documenting these Roshans and Mashrabiyahs and their com-ponents via the most recent digital approaches and the application of suitable repair methods will ensure longevity of preservation, and conservation of value and authenticity (Adas, 2013). Besides, the 3-D models of these important objects can be used to guarantee any future projects in the area that can preserve the authenticity of these projects.

Before modelling the Roshan and Mashrabiyah, it is very important to understand the architectural and structural rules associated with these elements. Through analysing the Roshan and Mashrabiyah, it can be seen that they are shaped via three main parts which are the head, the body, and the base. Next, each part is divided into three sub-types, for example, the head was divided into the crown or "Tajj" in Arabic (Figure 4.14 shows the main parts of the Roshan). By dividing the elements into sub-types, as has been done in the case of the Roshan, one can reduce the complexity and make the element easier to understand. Figure 4.15 shows the Roshan parts' layout.

Fourthly, each single part of the Roshan's structure was created and modelled, as close to its real condition as possible. Finally, after the 3-D modelling stage of the main structural, ornamental, and complex parts, this layered data must be

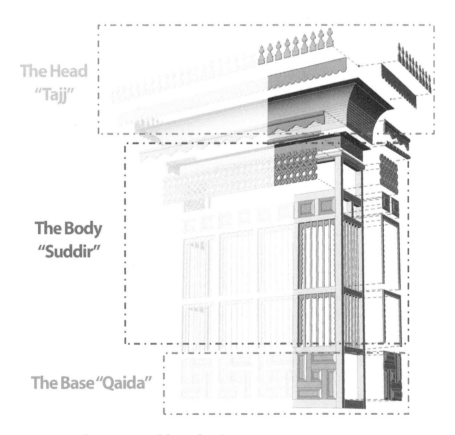

Figure 4.14 The main parts of the Roshan Source

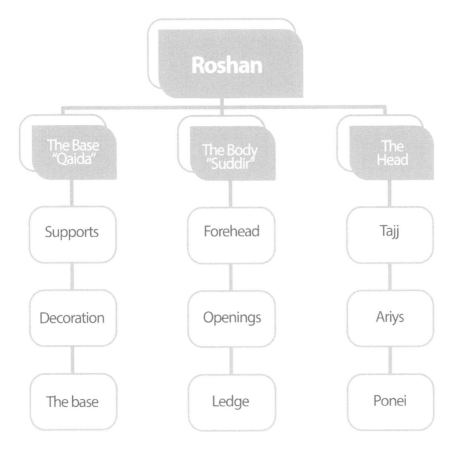

Figure 4.15 The Roshan parts' layout

integrated in order to complete the final 3-D model of the Roshan element, which will be appropriate for exportation to the JHBIM system. This model will then be appropriate for exportation to the JHBIM system.

Figure 4.16 shows the Roshan frame structure modelling. Figure 4.17 shows the steps in the modelling of the Roshan, and these objects will be used as a plug-in extension for existing BIM program platforms (see Figure 4.18). In this case, Autodesk Revit (2014, 2015, 2016, 2017 versions) has been used.

Figure 4.19 shows the steps of the Roshan modelling in "as-built" level of details.

4.4.2.3 Modelling the Manjur pattern, wooden shish, and woodwork

The Manjur pattern can be described as lattice grilles of wood or a shish net, which are always on the top part of the Roshan or the Mashrabiyah. The main aims of

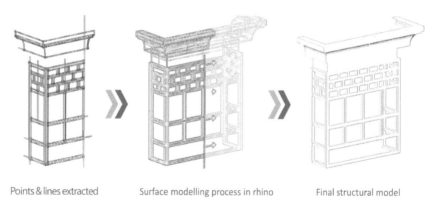

Points & lines extracted   Surface modelling process in rhino   Final structural model

Figure 4.16 The Roshan frame structure modelling

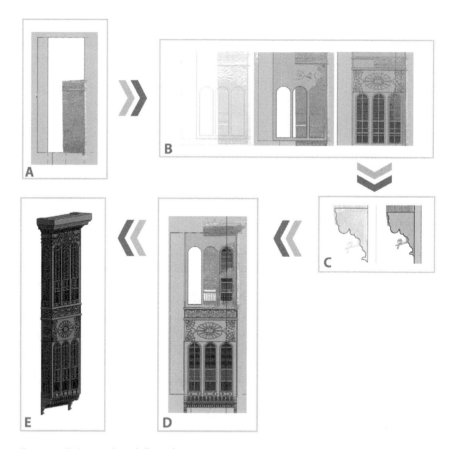

Figure 4.17 Steps of modelling the main Roshan of Nasif Historical House. (A) Inserting the cloud point into Revit. (B) The process of modelling the Roshan. (C) Modelling the base of the Roshan. (D) Modelling the upper part of the Roshan. (E) The final result of the 3-D modelling Point into Revit

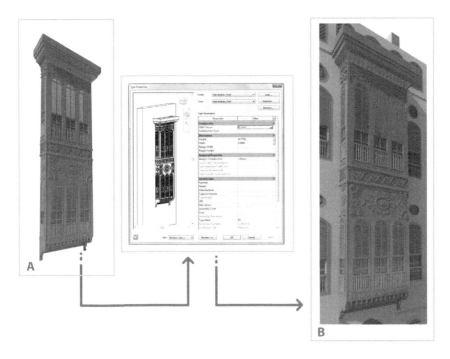

Figure 4.18 Inserting the main Roshan into the BIM model

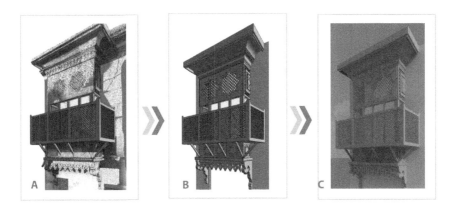

Figure 4.19 The steps of the Roshan modelling in "as-built" level of details

the Manjur are to allow gentle light in and to maintain shade, as well as allow a cool breeze to enter the building, which is desirable in the hot climate of Jeddah city. From a cultural point of view, the Manjur provides a veil, which permits the families inside the house to look outside without being seen (SCTA, 2013).

The concept of the beautiful Manjur designs is to make specially cut laths of wood fit into one another at right angles (criss-crossed) and set within a frame. These Manjur patterns offer a pleasant view from the outside, as well as from the inside. The form in which the sides of the laths are cut determines the form of the resulting open spaces in between, as well as the complete pattern of the Manjur net.

According to SCTA (2013), "The shish normally contains two or more shapes arranged in sequences so as to give the Manjur the desired pattern", and generally, the sizes and shapes of the Manjur patterns are "selected in such a way as to provide a balanced combination of shade, delicate light, nice breeze, as well as privacy".

The modelling process for the Manjur was based on the laser scanning data. The data was exported in (.rcp) format and inserted into Autodesk Revit in a separate family file. This step is very important because if this object was modelled in the same Revit project file, it is impossible for it to be used as a family block in another project.

The second step was understanding the architectural rules of the Manjur. The main concept of the attractive Manjur designs is to make specially cut laths of wood that fit into one another at the right angles (criss-crossed) and set within a frame. These Manjur patterns offer a pleasant view from the outside as well as the inside. The way in which the sides of the laths are cut determines the form of the resulting open spaces in between, as well as the complete pattern of the Manjur net.

The third step is modelling the lath and then duplicating this lath as necessary. Next, modelling the frame is required. Then, arranging the laths in the frame (Figure 4.20) and installing the object into the BIM model (Figure 4.21).

4.4.2.4   *Modelling the plaster decoration works*

The humid and salty weather in Jeddah affects the limestone and coral building blocks. As a result, the main solution the old builders used for protecting the walls and surfaces of these building is using plaster. Over time, the plasterer's craft has improved and developed, which has resulted in decorative plaster carvings on the façades. As can be easily observed, the decorative plaster is always centred on the ground level of the façades, particularly around the main windows and main doors of the Jeddah houses.

In relation to the process of making the plaster and the decoration, the work was applied to the coral walls when they were still wet. The beauty of the plaster decoration lies in the contrast between the engraved and non-engraved surfaces, which create a difference in the light and shadow levels. Great examples of the carved plaster decoration can be found in Jokhdar Historical House and Ribat al-Khonji.

According to SCTA (2013), "Though there is no scientific study devoted to the plaster decorative patterns in Historic Jeddah, and the Gate to Makkah, it appears that older decorations were simpler and more geometric, while later

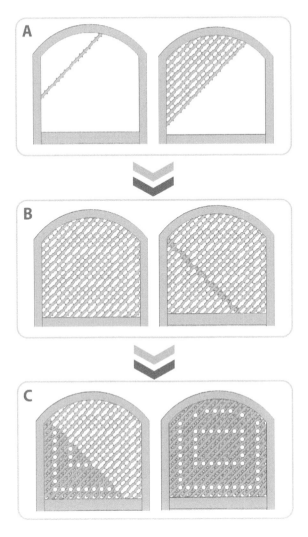

Figure 4.20 Modelling the Manjur pattern in 3-D. (A) Modelling the first part of the Manjur. (B) Combining the Manjur parts. (C) The final result of the 3-D modelling

carved plasters became more elaborated with complex floral decorations cut deeper in the plaster though remaining within its thickness".

An interesting aspect that has been found in some main façades of the buildings in Historic Jeddah is the sgraffito. The sgraffito has always been on the corner of the ground level of two façades and this sgraffito is created by scratching through

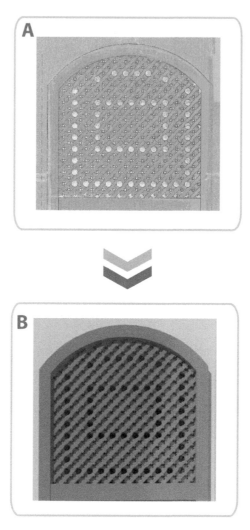

Figure 4.21 (A) Comparing the point cloud and the 3-D model of the Manjur. (B) Inserting the object into the model

a surface to reveal a lower layer of contrasting colour. These sgraffito are usually
in rectangular or square panels, yet, they rarely come in frieze patterns such as the
patterns on the Nasif Historical House.

In the case study of the Nasif Historical House, the style of the plaster decoration is floral. This decorative plaster is concentrated around the main gate of the
house and acts as a separator belt between the floors (Figure 4.22).

Moreover, to model such complicated patterns with many small details, close-
range photogrammetry was used alongside the laser scanning. The analysis

Figure 4.22 The plaster decoration on the façade of Nasif House

uncovered a lot of damage to this decorative plaster and therefore sections with less damage were chosen to be modelled.

The next step was to transfer the 3-D point cloud (Figure 4.24) to 3-D triangle meshes or polygon mesh surfaces via the MeshLab software (Figure 4.23). Afterwards, it is possible to export the 3-D Mesh to DXF file to be modelled in CAD (Figure 4.25). Then, it is possible to insert these (meshes) into the Revit model.

Furthermore, the triangle meshes, or polygon meshes, can be segmented and created from the point cloud and image data. The resulting meshes can be utilised

Figure 4.23 The plaster point cloud converted to a 3-D Mesh surface via MeshLab

Figure 4.24 The 3-D plaster point cloud

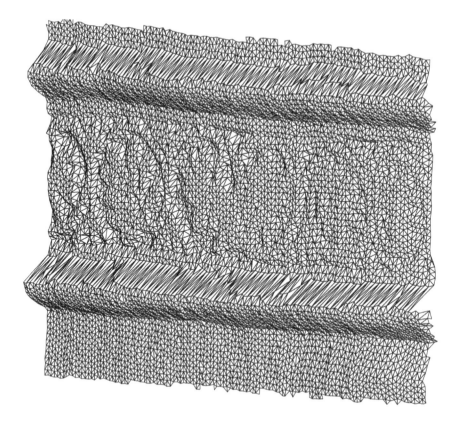

Figure 4.25 Exporting the 3-D Mesh surfaces as a DXF file

for a singular purpose or reused where they appear again in other buildings in the historic district of Jeddah.

4.4.2.5   *Modelling the main gates and the doors*

Another important architectural characteristic in historic Jeddah are the doors, both external and internal. What is more, these wooden doors, according to SCTA (2013), "have double leaves and are decorated with carved panels representing some of the finest carpentry and decoration in Arabia". The main entrance doors are given considerable attention in the traditional houses of Jeddah city.

Regarding the design of these entrance doors, it can be noted that these doors are quite tall, have elaborately decorated panels, carved designs on both sides, and are crowned with carved stone or decorated plasterwork. According to SCTA (2013), these designs "are in the form of repeated floral motifs and rosettes linked by geometrical patterns and/or multi-sided polygons or pointed stars". Also, it can be noted that in some parts of the doors, carving designs are shallow and deeper in others.

For the modelling of the gates and the doors of the case study, the first step was capturing the details of the gates and doors based on the laser scanning and close-range photogrammetry. The second step was to export the laser scanning point cloud to (.rcp) format to be modelled in Autodesk Revit as a family block. The third step was to insert the modelled gate and door into the Nasif House Revit project. Figure 4.26 shows the steps of modelling the main gates of Nasif House. Figure 4.27

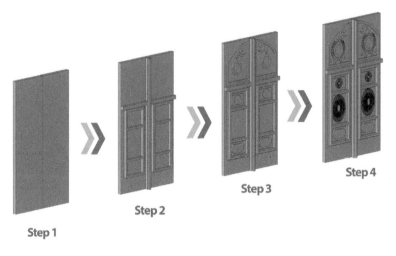

Step 1

Step 2

Step 3

Step 4

Figure 4.26 The modelling steps of the main gate

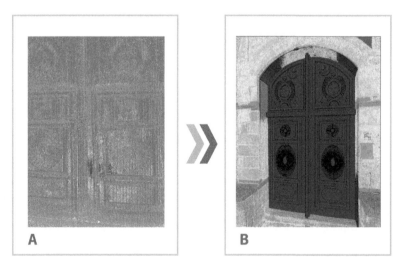

A

B

Figure 4.27 Modelling the main gates of Nasif House. (A) The point cloud of the main gate. (B) Modelling the gate onto the point cloud

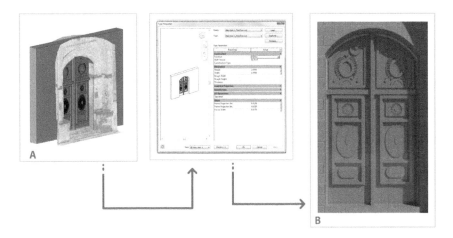

Figure 4.28 Inserting the main gate into the BIM model

and Figure 4.28 show inserting the object into the BIM model of Nasif House. Figure 4.29 shows the steps of modelling the internal door of Nasif House.

4.4.2.6   Modelling the penthouse and the "Tarmah"

The penthouse, or the "Tarmah" in Arabic, is the most important part of Nasif Historical House. This is because it is located in the highest place of the house and overlooks all of the historic district (Figure 4.30, Figure 4.31).

The modelling of the Tarmah started with point cloud data. The generated 3-D point cloud data of the Tarmah was too heavy due to the huge detail and the high resolution of the laser scanning. This caused many issues and program clashes. To solve this issue, the 3-D point cloud of the Tarmah was divided into four parts and each part presented one of the elevations. Afterwards, these parts were combined into one model and inserted into the Nasif House project. Figure 4.32 shows the modelling steps of the Tarmah. Figure 4.33 shows the process to insert the object into the BIM model. Figure 4.34 show the steps of modelling the Hijazi window.

The case study includes a lot of windows and Roshan types. Each of these objects has their own characteristics. However, they are produced, following a common procedure and hence follow a standardised form. This can make the modelling slightly easier to be understood (see Figure 4.34 to Figure 4.44). However, the difficult part can be in the LoD of the window decorations.

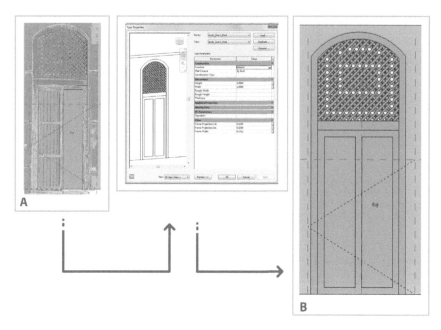

Figure 4.29 Modelling the internal door of Nasif House

Figure 4.30 Tarmah from the outside
Source: Author

Figure 4.31 The view from one of the Tarmah's windows
Source: Author

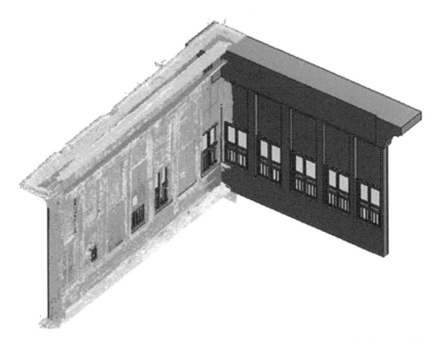

Figure 4.32 The steps of modelling the Tarmah. (A) Inserting the point cloud and modelling the first part. (B) The complete of the 3-D model of Tarmah

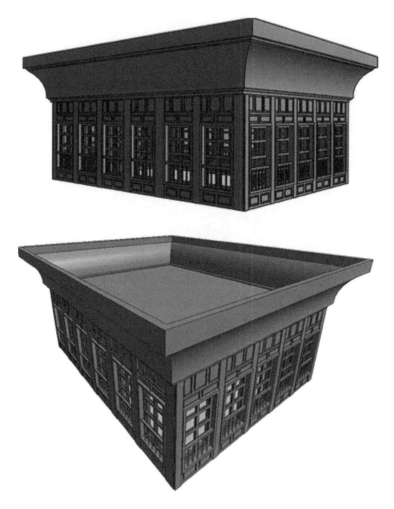

Figure 4.32 (Continued)

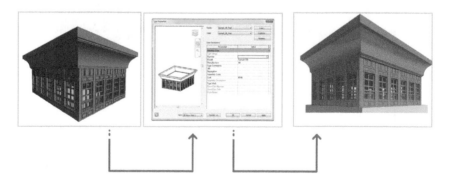

Figure 4.33 Inserting the Tarmah Revit family block into the model

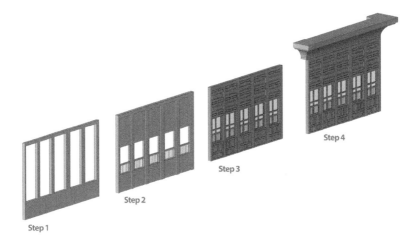

Step 1

Step 2

Step 3

Step 4

Figure 4.34 The steps of modelling the Hijazi window

A

B

C

Figure 4.35 Modelling the side Roshan of Nasif Historical House

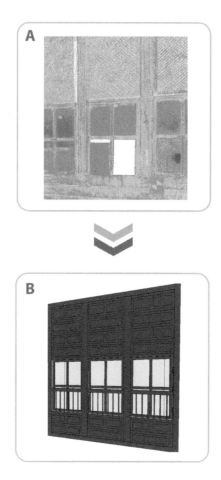

Figure 4.36 Modelling the side window of Nasif Historical House

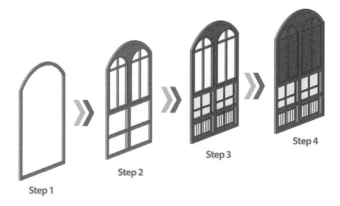

Figure 4.37 The steps of modelling the window as Revit family

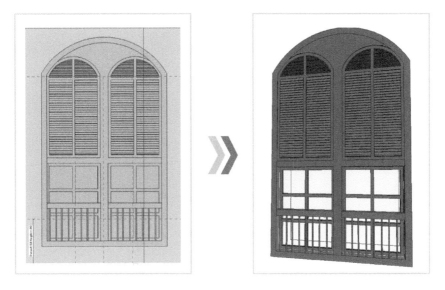

Figure 4.38 Modelling the main elevation window of Nasif Historical House

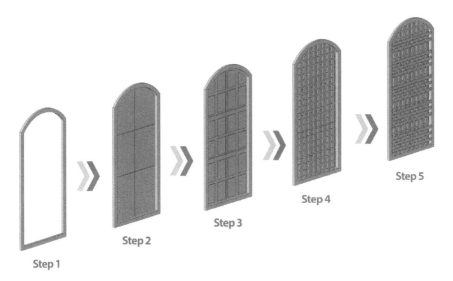

Figure 4.39 Modelling one of the Hijazi windows with Manjur pattern in "as-built" level of details

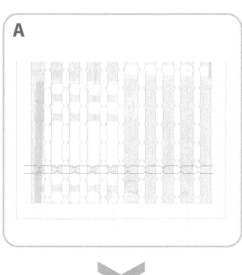

Figure 4.40 Modelling the Manjur for the window in 3-D

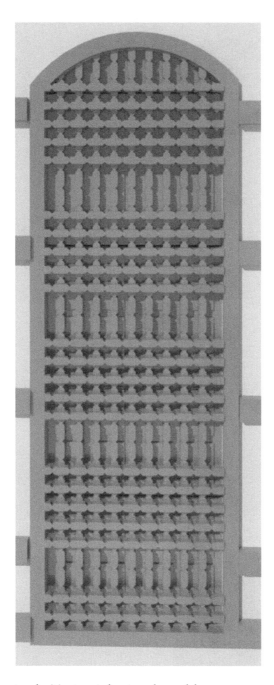

Figure 4.41 Inserting the Manjur window into the model

Figure 4.42 Modifying the object to create another (window type 1)

Figure 4.43 Modifying the object to create another (window type 2)

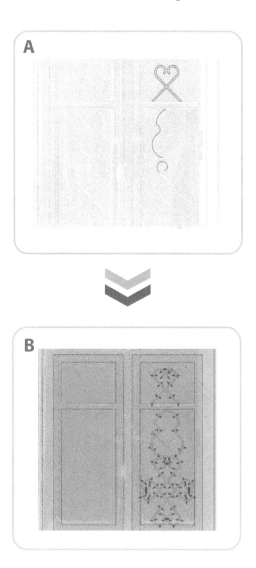

Figure 4.44 Modelling one of the windows in "as-built" level of details

4.5 The case study—Nasif Historical House

4.5.1 *Cultural significance of the case study*

Nasif Historical House is one of the most famous houses in the historic district. It was built at the end of the 18th century by Sheikh Omar Afandi Nasif, the agent of Jeddah for the Sharif of Makkah (Figure 4.46).

The main reason for the house's reputation is during the time when King Abdul-Aziz (Figure 4.45) entered Jeddah in 1925, he stayed in Nasif House, where

Figure 4.45 King Abdul-Aziz Al Saud

Figure 4.46 Sheikh Omar Afandi Nasif

he met the most notable people of Jeddah, as the house was supposed to be the most distinguished and appropriate residence for the Sultan.

The architectural style of Nasif House is similar to other historic Jeddah buildings with Roshans and Mashrabiyahs.

Until 1975, the house belonged to the Nasif family. One of the heirs, Sheikh Muhammad, turned Nasif House into a private library containing more than 16,000 books.

Today, Bayt Nasif has been restored and has become a museum and cultural centre (SCTA, 2013).

4.5.2   Nasif Historical House description

Nasif House was constructed at the end of the 19th century (the 13th century, according to the Hijri calendar). The house was constructed by Sheikh Omar Afandi Nasif, who started the building work in 1872 A.D. (1289 Hijri) and finished in 1881 A.D. (1298 Hijri). The man in charge of construction, Sheikh Omar Afandi Nasif, was a well-known person in the city. The Sultan of the Ottoman Empire, based in the city of Istanbul at the time, also rewarded Sheikh Omar Afandi Nasif with the award of "Vizier", as well as the awarding of the level of Mecca's Ameer's (Sharif "Awn") aide or spokesperson. The Nasif House is a clear representation of the stature of the person who built and owned it. The Nasif House also manages to capture the pinnacle of construction and building architecture at that point in time.

Nasif House comprises four floors and has an architectural design that has identical dimensions, giving the Nasif House a very pleasing and aesthetic look (Figure 4.47). The Nasif House is situated right in the middle of the town. The external façade of the building is thoughtfully constructed with aesthetically created corners. However, the main attraction of the whole building is the beautiful woodcarvings on the main door, windows, and other wooden features. This carving has been done in impeccable quality wood that was brought into the city from eastern countries (Nawwar, 2013).

There are two doors on the ground floor of the building at the main entrance. These doors have been divided into smaller segments and beautiful woodcarving has been done onto each separate segment (Figure 4.48).

The vertical windows have been decorated with traditional glass work called the Mashrabiyah and are accompanied by shutters. The colour of the woodwork in the building is further enhanced by the faint colour of the building walls, providing a beautiful contrasting view. Apart from the fact that these windows have been aesthetically designed with meticulous woodwork, the louvered shutters also serve an important purpose as they allow for proper ventilation of the building, while also providing privacy for the occupants. When someone is standing in the front of the building, he or she is bound to feel over-awed by the Roshan that spans over two floors, starting from the main entrance gate and extending upwards. Alongside the large Roshan are the windows of the fourth-story rooms with their

Figure 4.47 Nasif Historical House, Historic Jeddah, Saudi Arabia
Source: Author

beautiful, aesthetically pleasing, and meticulously carved woodwork. When viewed in conjunction, the Roshan and windows are a sight to behold. The front façade of the building is also further enhanced by careful carvings on the wall of the building. These carvings serve to further support the beauty of the building rather than trying to stand out on their own (Nawwar, 2013).

Figure 4.48 The main door and the Roshan of Nasif House
Source: Author

Figure 4.49 shows the site plans of Nasif House. The inside of the building has also been carefully thought out and developed with big rooms in the centre, flanked by smaller living rooms on either side. The large twin flight of stairs is situated in the middle area of the back part of the building. The rooms for the helpers and servants are also situated at the back of the building. The interior of the building stands out just like the exterior in the sense that every storey presents a unique artistic construction. For example, the first storey houses a unique arch that is angular in nature, along with a "Diwan" (an Arabic word that means the main hall). The second storey has a very beautifully designed place in the lobby, while the third storey boasts a bathroom with a little dome in the ceiling, known as the Turkish bath. The fourth storey offers a huge terrace that extends in front of the central room, along with a kitchen situated at the back of this storey.

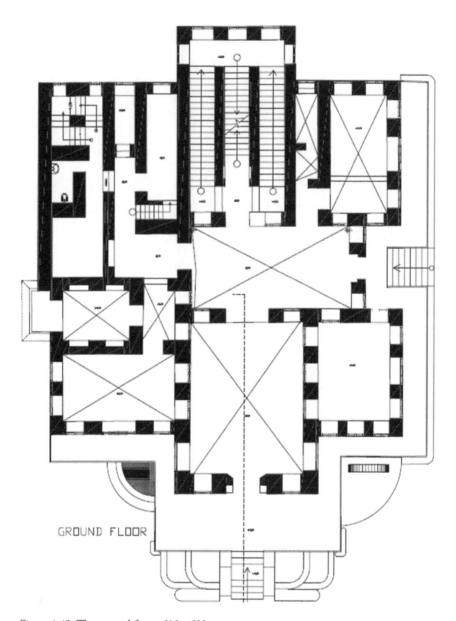

GROUND FLOOR

Figure 4.49 The ground floor of Nasif House

Finally, the penthouse is constructed with fine woodcarvings that share their designs with those situated on the floor just below the penthouse.

Coral rocks extracted from the bed of the Red Sea have been carved in shapes with equal length and breadth before being used to construct the walls of the

Figure 4.50 A coral block or in Arabic "Mangabi"
Source: Author

building (Figure 4.50). Plaster is used to smoothen out the look of the construction, while wood pieces are used at uniform distances in construction to lend further strength to the building's structure. Although the wood pieces have been covered with plaster from the outside to give a smooth and consistent look, the insides have been left without plaster.

The ground of every floor has been constructed using wooden beams. Mats made of palm trees have been used, along with small stones and earth, in order to smooth out and harden the flooring. Finally, to finish the flooring, screed has been poured over the floor. The eastern area of the south of the building features a utilities area. In the building, there are two floors that were built primarily to house different utilities, such as toilets, stairs, storerooms, kitchens, and so on.

The eastern face of the building shows a clear fissure in the wall that divides the area of residence for the owners of the building from the area that housed the servants. The rear face of the building also presents a unique look as that has been left without plaster so that the horizontal wood boards, along with the vertical windows, present a very asymmetrical yet aesthetic look. The penthouse of the Nasif House is situated in the very middle of the roof and is adorned with beautiful carvings that share designs with the carvings done on the faces of the corner rooms on the fourth storey. One of Muhammad Nasif's sons distinctly recalls that the penthouse was his grandfather's favourite place, and this is where he used to sleep most often. If someone enters the Nasif House and passes by the reception, he would immediately enter the huge main chamber of the house, opposite to which is an angular arch. The large chamber here was used to house the vast library of Muhammad Nasif until it was shifted to the King Abdul Aziz University (Nawwar, 2013). The staircase at the Nasif House is so colossal that it was used in the past as a route for water carrying camels to reach the top floor. One of the grandsons of Muhammad Nasif remembers how his grandfather used to take his horse up to the top floors simply by riding the horse up the giant staircase.

The Nasif House is also home to some beautiful pieces and designs from Islamic architecture as well. For example, the Mashrabiyah are used all over this iconic building, from the external walls to the interior windows. There was even a storage facility for collecting rainwater during times of rain. During times of no rain, water was delivered to the house by water-carrying camels, who brought water from wells situated on the outskirts of the city.

Currently, the ground floor of the building has been turned into a museum by the municipal commission of Jeddah. Moreover, the commission also seeks to convert the whole building into a museum.

4.5.3 Technology, programs, and equipment used

Nasif Historical House is unique and complex as it has a large amount of Hijazi architectural detail. In order to build the model and coordinate it through the information for the historic house, the 3-D laser scanner has to generate 3-D point clouds, which forms the basis of the modelling process. This laser scanning procedure has many features, such as a field of view of 360° in a horizontal direction and 270° in a vertical direction, and this feature allows for the capturing of full panoramic views (Leica Geosystems, 2013). Another feature of the scan method is described as a "time of flight dimension principle", which means that the Leica Scan-Station C10 can scan at ranges between 2 m up to 300 m (Figure 4.51). The laser scanner has a mixed pixel clarifying framework, thus creating highly accurate point clouds which have x, y, and z coordinates. The Scan-Station C10 resolution and the spot size ranges from 0–50 m to 4.5 mm (FWHH-based) and 7 mm for (Gaussian-based). The accuracy of the single measurement for the position is 6 mm and for the distance is 4 mm at a 1-m to 50-m range. As with many modern laser scanning techniques, the Leica C10 is provided with an internal HD camera. The scanner is likewise good with GPS, thus can organise the fieldwork with an encompassing ecological setting. Besides, through the "Truview" feature, the LiDAR images can be presented, which provides the ability to measure the building's dimensions.

In regards to the case study, to process the laser scanning point cloud, both Leica Cyclone version 8.1.1 and Autodesk Recap Pro software were used for the registration process and removing noise from the laser scanning data. This 3-D data was then delivered to Autodesk Revit to be modelled. Moreover, AutoCAD 2016 was used to produce the Ortho drawings based on the point cloud data. In addition, for the image survey step, a professional Nikon D5100, 18 mega-pixel camera with an

Figure 4.51 Leica laser scanning C10

Source: Author

18–55 mm f/3.5–5.6 autofocus lens was used to capture the building's architectural features and details. The 3-D JHBIM model was built using Autodesk Revit 2015. Finally, Adobe Photoshop CS5® was used to produce the images.

4.6   JHBIM on-site data acquisition

4.6.1   *Images survey*

To complete the JHBIM project, the on-site work began with the image survey step, which was intended to highlight and document the Hijazi architectural features for the Nasif Historical House. The image survey of the house involved ten working days (from the 12th to the 25th of August 2014), and highlighted that the Nasif Historical House has unique architectural features related to different Islamic cultures and ages. These elements included the Roshans and Mashrabi-yahs, which were borrowed from the Ottoman culture.

Through using professional software, such as Autodesk Recap, Autodesk memento, and Agisoft Photo-scan, it is possible to use these images to build the 3-D mesh of the object. This process is known as architectural photogrammetry (for more, see section 2.5.6.4).

4.6.2   *Terrestrial laser scanning survey*

The concept of remote data capture using terrestrial laser scanning, in addition to the processing of the point cloud data (previously explained), is employed. This concept formed the basis of the initial elements for the JHBIM procedure. This procedure describes data acquisition based on the TLS and is followed by the data processing of point cloud data.

Employing the terrestrial laser scanning method involved initial preparation of the location in order to indicate the scan-station points to gain a good overlap between these scan-stations. Further, this was done to achieve high-quality scanning, thus the HDS White and Black 'W/B' targets method was employed and about 400 'W/B' targets were used (Figure 4.52 shows using the W/B targets in the staircase).

The preparation of the location lasted for more than five working days to obtain a successful laser scanning survey. Especially in the case of historical sites, it is very important to select the most suitable viewing locations, particularly since the amount of achievable scan-stations is normally limited due to the complexity of the structure and site.

Moreover, locating the targets on the staircase walls took more time as the house staircase targets, in this case, were used to register all the floors scans (Figure 4.53 shows the registration by using the staircase).

After preparing the location, the next step involved the laser scanning survey, and to obtain high accuracy in the laser survey, at least three (W/H) targets were needed to be in common with each of the overlapping scan-stations (Figure 4.54).

Figure 4.52 Using the W/B targets in the staircase

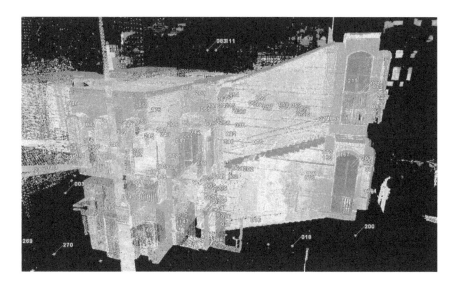

Figure 4.53 Using the staircase to link the floors' scans and to register the scans

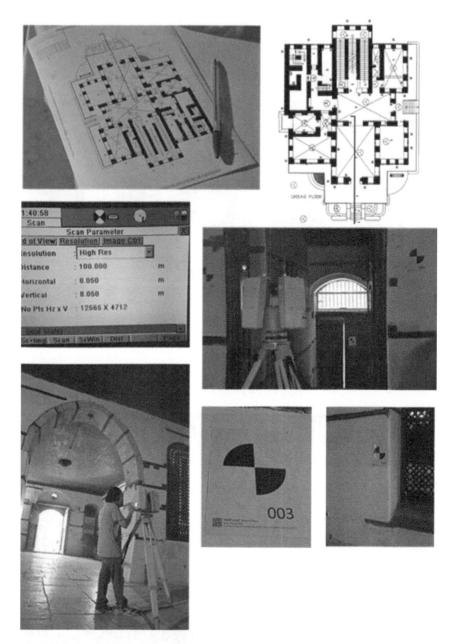

Figure 4.54 Terrestrial laser scanning survey process

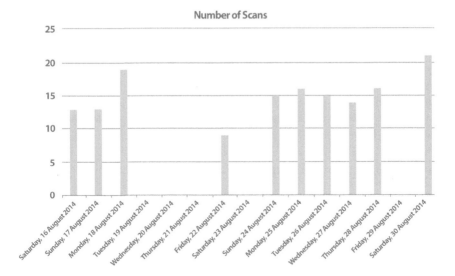

Figure 4.55 Graph illustrating the number of scans per day
Source: Author

The scan-stations involved about 150 scans (overall, about 112 hours), which produced more than 70 GB of point cloud data with a resolution of 0.05 m and an average of 10 m distance inside the house, and up to 30 m outdoors.

The laser-scanning step took over ten working days to be completed (from the 16th to the 30th of August 2014) (see Figure 4.55).

The framework of the laser scanning survey is based on the following steps.

The first step: exploring the site and taking notes.
The second step: preparing the site and determining the scan locations and the W/B targets' locations.
The third step: starting to collect the data using the Leica scan-station C10 from inside to outside the house.
The fourth step: transferring the data to the desktop computer and preparing them for the next stage; "the point cloud process and registration".

4.6.3 *Terrestrial laser scanning data processing*

Following the data capture, the LiDAR model develops over several of processing steps involving registering different scans, cleaning, and organising this point cloud data. The required 3-D point cloud models can be generated by using a number of software programs, such as Polyworks, Autodesk Recap, and Leica Cyclone.

4.6.3.1   *Point cloud registration*

In this case study, Leica Cyclone 8.1.1 was initially used to register the scans. Several Scan-Worlds corresponding points in overlapping sections were linked for registration purposes. The point sets were registered in accordance with the coordination system and the elevation above sea level. This step, according to Attar et al. (2010, p. 177), "allows for geo-referencing of the entire dataset by knowing the precise position of each point in terms of real-world coordinates".

Figure 4.56 shows the section on the 3-D point cloud after using the staircase to link the floors' scans and to register the scans.

4.6.3.2   *Point cloud data cleaning*

After the registration step, the Leica Cyclone was used to remove noise from the registered 3-D point cloud (Figure 4.57), and completion of data processing required a total of ten working days.

Figure 4.58 to Figure 4.62 show the resulting combinations of scans for the 3-D point cloud model, and the average resolution of the 3-D point cloud model of Nasif Historical House was 70 mm on the object surface.

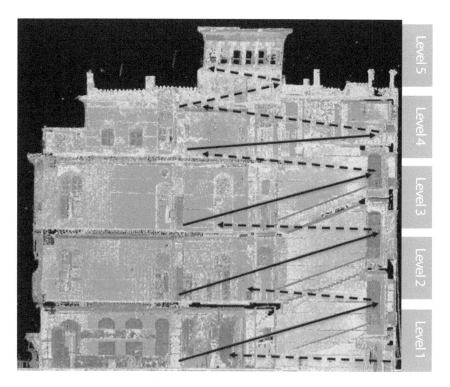

Figure 4.56 Section on the 3-D point cloud after using the staircase to link the floors' scans and to register the scans

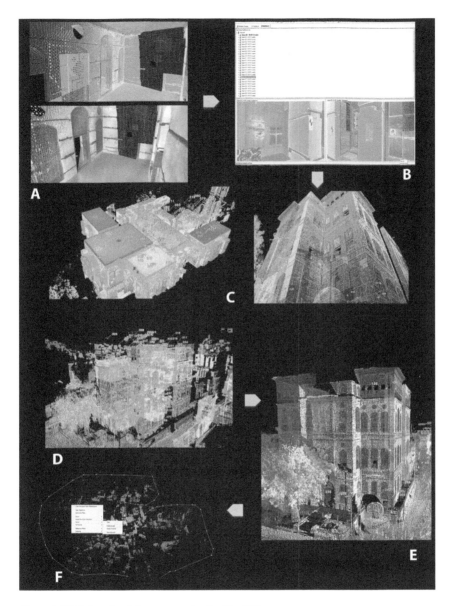

Figure 4.57 Point cloud data processing for Nasif House

Summary of the required output information for the JHBIM procedure are:

1   A completed 3-D point cloud model of the whole building in (.rcp) file formats, indicating the plans, the elevations, structures "as it can be noticed", and the 3-D model.

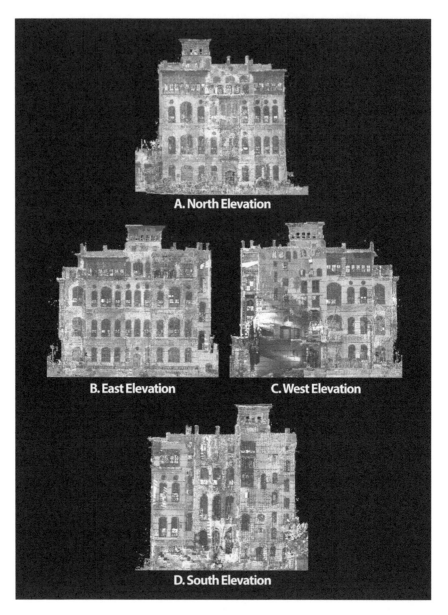

Figure 4.58 The resulting combinations of scans for the 3-D point cloud model. (A) North elevation. (B) East elevation. (C) West elevation. (D) South elevation

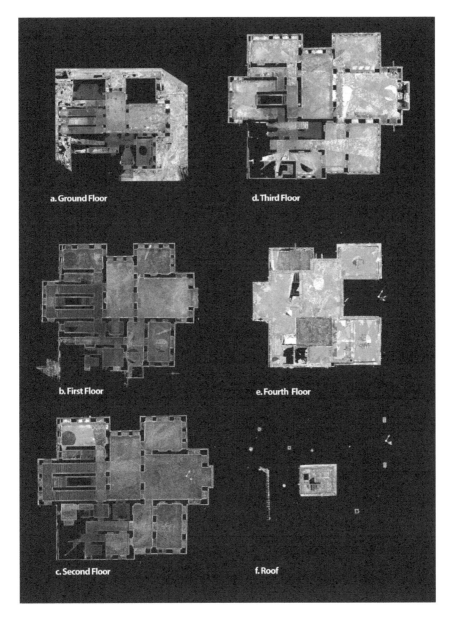

Figure 4.59 The resulting combinations of scans for the 3-D point cloud model

Figure 4.60 3-D point cloud of Nasif Historical House in Cyclone

Figure 4.61 3-D point cloud of Nasif Historical House

Figure 4.62 The laser scanning of the main Roshan in Nasif House

2   Providing the architectural elements of the heritage building, which is unparalleled in the architectural libraries in (.rcp) file formats to be modelled separately and to be stored in the Hijazi architectural objects library, "HAOL".

4.7   The JHBIM modelling process

The purpose of using HBIM for historical Jeddah buildings was to develop an interactive model to move from the basic level of BIM "CAD and 2-D drawings" toward more advanced levels of BIM "level 2 and level 3" (Baik et al., 2014). Additionally, another goal was improving the productivity of component modelling while reducing manpower input. Conventional methods either extract highly qualitative mechanisms just within perfect conditions (Budroni and Böhm, 2010) or directly generate the 3-D model with lots of mistakes (Xiong et al., 2013). Moreover, the use of HBIM also supports the preservation and sharing of information in relation to these historical buildings with experts and others involved in the decision-making processes (Eastman et al., 2011; Fussel et al., 2009). Saygi et al. (2013, p. 283) stated that, "BIM will provide the possibility to represent all views (3D model, plans, sections, elevations, and details) automatically". These findings are supported by Murphy et al. (2009), who suggested that the purpose of employing BIM in the historic field is to offer interactive parametric objects that represent architectural elements to model historical monuments for the TLS point cloud.

The first step toward JHBIM was to reduce the complexity of the historical building and complete the project. Therefore, the Hijazi Architectural Objects Library (HAOL) used LiDAR point cloud data, image survey, and Hijazi architectural pattern books (Baik et al., 2014) to reduce the modelling time or project delivery time (PDT) and further, provide a high level of detail (LoD). The framework for building the HAOL for the Nasif Historical House project began by understanding and analysing the architectural manufacturer's rules regarding the house's objects. Then, classifying the Hijazi architectural objects based on several standards, such as the amount of detail, the style, the shape, and the similarity between these objects (see section 4.4).

The second step is to model the heritage building structure. Moreover, in the case of Nasif Historical House, Autodesk Revit was used as it provided quick modelling and allowed for changes to the 3-D model, as well as a high quality of construction documents and a high level of flexibility. The modelling step of the JHBIM model of Nasif Historic House began once the 3-D point cloud was received. Some parts of the 3-D point cloud, such as the Roshan and the Mashrabiyah, were selected to be modelled separately for the HOAL library. These elements were inserted into the model later (see section 4.4.2.2), while the building's main structure was modelled based on the 3-D point cloud (Figure 4.63 shows preparing the elevation "in point cloud format" in Revit with the levels to be modelled).

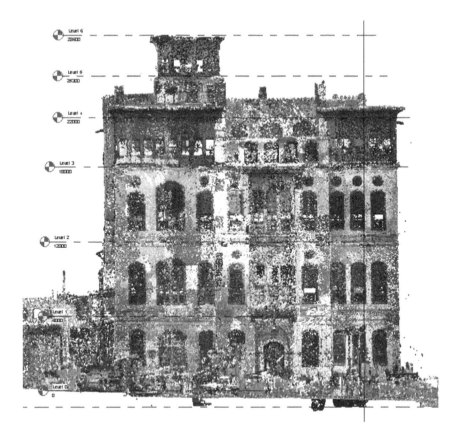

Figure 4.63 Preparing the elevation "in point cloud format" in Revit with the levels to be modelled

In the case of the Nasif Historical House project, the modelling step started with building the house's external and internal walls of the lower floors and then the upper floors (Figure 4.64, Figure 4.65, Figure 4.66, and Figure 4.67). Next, the façades were simplified as much as possible to be modelled (see Figure 4.68). In this step, it was very important to build the house by using the traditional methods of construction that were used to build the house, in order to allow us to calculate the loads on the walls and the floors for any future use.

The third step, which involved using the HAOL as a plug-in for the model, saved a lot of time and work. For example, instead of wasting time creating each single part of the Roshan window, one form of the Roshan window was used, and the ability to modify and to scale the Roshan's detail saves a lot of time (Figure 4.69 shows modelling the window as a Revit family block to be inserted into the Nasif House project).

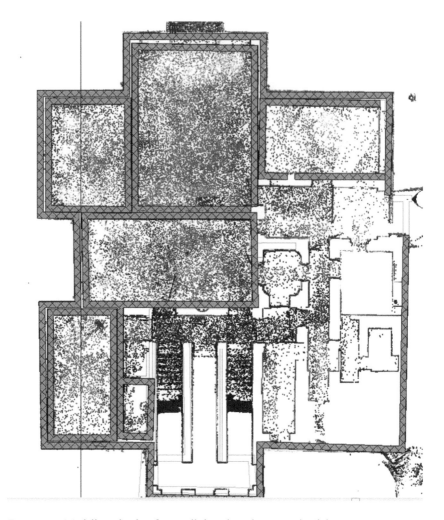

Figure 4.64 Modelling the first floor walls based on the point cloud data

However, some objects of the HAOL, "such as the windows, Roshan, and doors", were modified to fit exactly in the model. This is due to the fact that most of these elements had been thought over and designed on-site in order to fit into specific locations in the house (section 4.4.2 explains more about this).

The fourth step can be described as the makeup step for the BIM model, which requires adding the details that can provide a realistic sense to the model. These details are not construction elements and do not have any impact on the structure of the house; for example, the plaster decoration or stone endings on the walls which are known in Arabic as "Arais" (Figure 4.70). Figure 4.71 shows the 3-D modelling of the Arais for Nasif House.

Figure 4.65 Modelling the second level of the house (lighter areas show the extracted façades)

Figure 4.66 Modelling the third level of the house (lighter areas show the extracted façades)

Figure 4.67 Modelling the ground level of the house (lighter areas show the extracted façades)

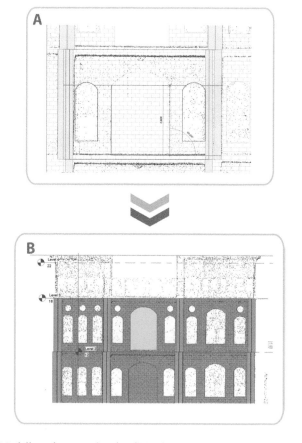

Figure 4.68 Modelling the main façade of Nasif House

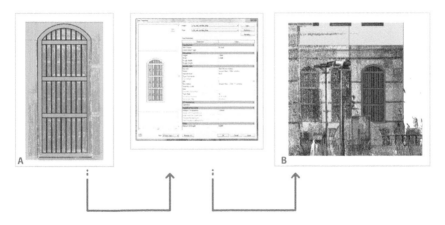

Figure 4.69 Modelling the window as a Revit family block to be inserted into the Nasif House project

Figure 4.70 An example of the Arais in Nasif House

Source: Author

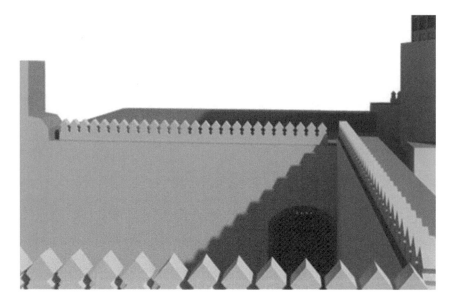

Figure 4.71 3-D modelling of the Arais for Nasif House

The final step is to add any available BIM layers such as "structures, ventilation, electricity, water, sewerage, and air conditioning". Some of this information originated from previous engineering work and reports that have been done over the last few years. Due to the nature of this data, it was added in manually.

The information and drawings of the JHBIM model are detailed and represented as 3-D architectural documentation; this information can help with meeting the UNESCO World Heritage designation with all the requirements they have for the nomination file (section 4.8 will explain more about this).

4.7.1 Inserting the HAOL into the JHBIM model

The Hijazi Architectural Objects Library (HAOL) has reproduced Hijazi elements as 3-D computer models, which are modelled using a Revit Family (RFA). The concept of this library is to use it as a plug-in for existing BIM software platforms such as Autodesk Revit. This plug-in can be accessed through the Add-Ins software ribbon, which is found in Autodesk Revit. Moreover, as such this plug-in can be developed through using Microsoft's Visual Studio. Through creating the Hijazi Architectural Objects Library, the delivery time for modelling the heritage building in Jeddah was reduced. Moreover, this library will support any future projects that feature in Hijazi architectural styles.

Figure 4.72 shows the Hijazi Architectural Objects Library (HAOL) inserted into the JHBIM model of Nasif Historical House.

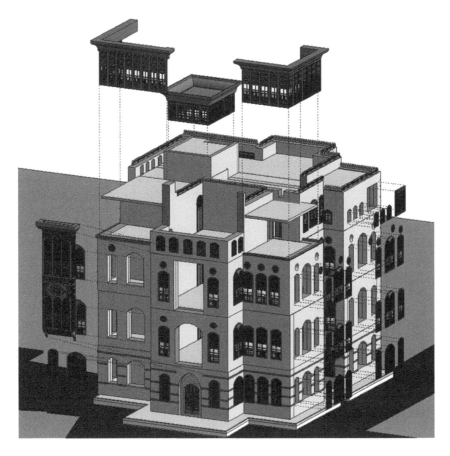

Figure 4.72 Inserting the HAOL into the JHBIM model of Nasif Historical House

4.8 Providing the engineering information

Using the BIM software platforms, such as Autodesk Revit, can provide an automatic way to produce fully engineering information such as elevations, plans, sections, working drawings, schedules, and animations and 3-D models. As such this engineering information can be the main source for any conservation and restoration projects in the World Heritage Sites. Furthermore, using BIM in the heritage field can generate comprehensive engineering information about the heritage site. Besides, to generate a complete HBIM model with a high level of detail, an object parametric library must be generated. This library should be used as a plug-in for the BIM software platform, which in this case is Autodesk Revit, and creating the objects library can be based on Revit family.

Regarding the case study of Historic Jeddah building, and through the integration of the BIM software platform and the point cloud data, it is possible to generate drawings based on the laser scanning data. For example, Figure 4.73 and Figure 4.74 show examples of the elevation and the section that can be produced from the point cloud via the BIM environment.

Moreover, through the JHBIM model, complete engineering information can be automatically produced. For example, Figure 4.75, Figure 4.76, Figure 4.77, and Figure 4.78 show the building elevations that were produced from the JHBIM model of Nasif Historical House. Figure 4.79 illustrates the floor sections of the ground floor plan produced from the JHBIM model. Examples of the detailed drawings of the Hijazi architectural object produced from the JHBIM model are shown in Figure 4.80 and Figure 4.81. Moreover, in Figure 4.82 and Figure 4.83 examples of the engineering information produced from the JHBIM model combining with the point cloud are illustrated.

In Figure 4.84, the result of the 3-D laser scanning of the Nasif Historical House and the BIM modelling for the heritage structure are presented; in Figure 4.85, sections of the walls, floors, and Roshan are displayed.

Lastly, at any location in the JHBIM model, the 3-D documentation can be created, which can be suitable for preservation implementation. For example, Figure 4.86, Figure 4.87, and Figure 4.88 show the 3-D JHBIM model of Nasif Historical House after the rendering in Autodesk Revit.

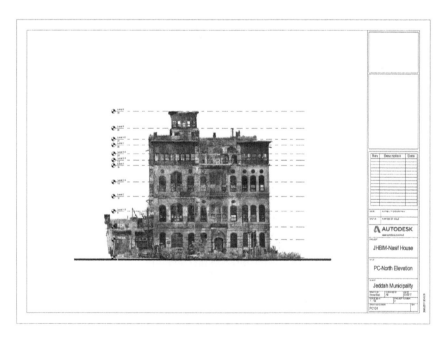

Figure 4.73 The north elevation of Nasif House in point cloud

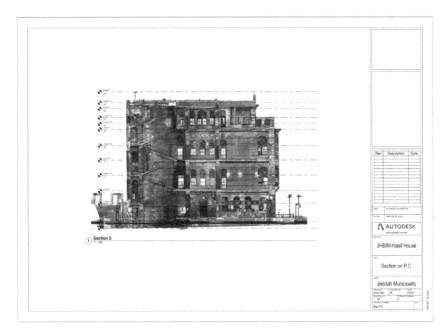

Figure 4.74 Section 1–1 on the point cloud

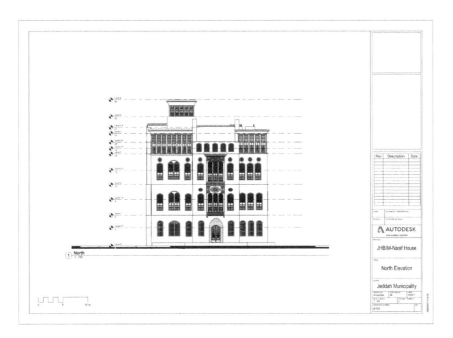

Figure 4.75 The CAD drawing of the north elevation of Nasif House

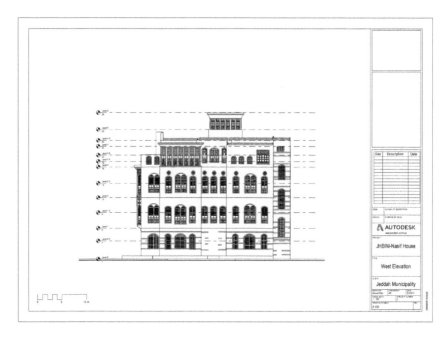

Figure 4.76 The CAD drawing of the west elevation of Nasif House

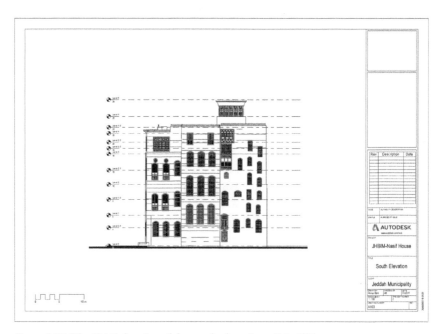

Figure 4.77 The CAD drawing of the south elevation of Nasif House

Figure 4.78 The CAD drawing of the east elevation of Nasif House

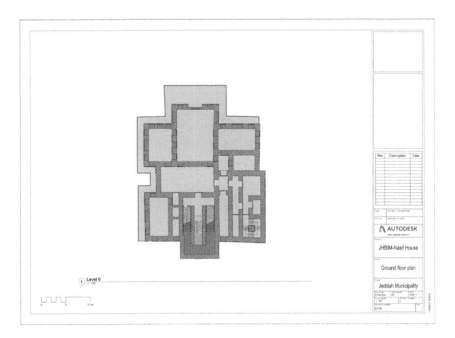

Figure 4.79 The ground floor plan produced from the BIM

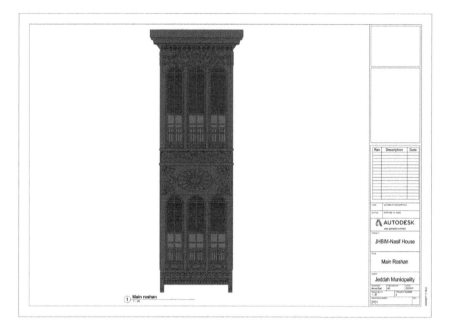

Figure 4.80 The main Roshan façade of Nasif House

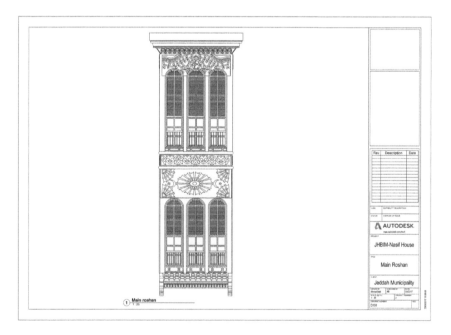

Figure 4.81 The main Roshan façade of Nasif House, CAD drawing

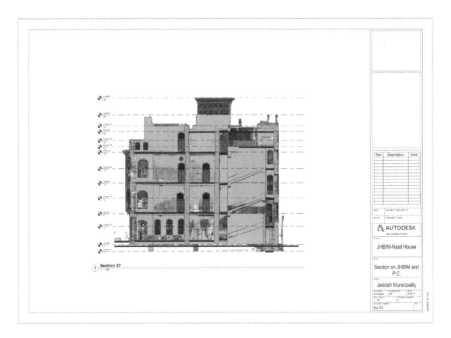

Figure 4.82 Section 1–1 on the JHBIM model and the point cloud

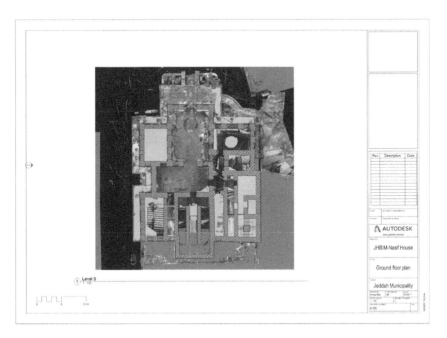

Figure 4.83 The ground floor plan produced from the BIM combining with the point cloud

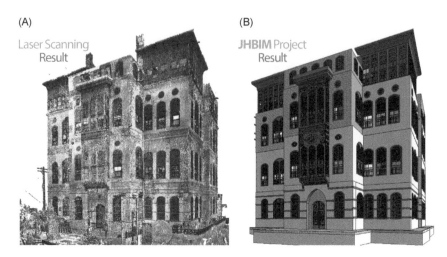

Figure 4.84 (A) The 3-D LiDAR point cloud of Nasif Historical House. (B) The 3-D model of Nasif Historical House based on JHBIM

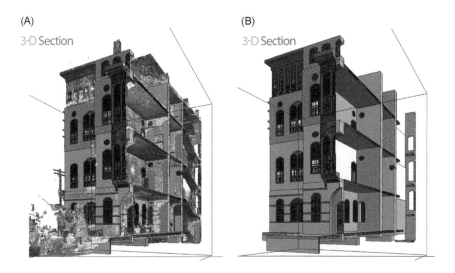

Figure 4.85 (A) 3-D section of the point cloud. (B) 3-D section of the JHBIM model

Figure 4.86 The 3-D JHBIM model of Nasif Historical House after the rendering in Autodesk Revit 2017. The northwest perspective

Figure 4.87 The 3-D JHBIM model of Nasif Historical House after the rendering in Autodesk Revit 2017. The northeast perspective

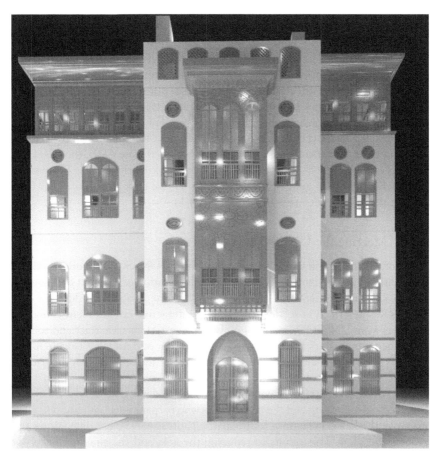

Figure 4.88 The 3-D JHBIM model of Nasif Historical House after the rendering in Autodesk Revit 2017. The north elevation perspective

4.9   JHBIM modelling time scale

During the pilot case study, the time modelling was around six months. This was due to lack of the digital architectural elements library being available in order to build the Heritage BIM model. Another issue was creating the architectural elements of the pilot case study that took a long time to be modelled. This issue was affected by the schedule of building the Jeddah Heritage BIM model of Radwan House (the pilot case study) (see Figure 4.89).

In the main case study of this thesis, through employing the Hijazi Architectural Objects Library (HAOL) as a plug-in Autodesk Revit family to the JHBIM model of Nasif Historical House (the case study) (see Figure 4.90), the use of this technique facilitated and sped up the process of creating the model. The process of modelling the case study of Nasif Historical House took six weeks rather than six months as the pilot case study did (Table 4.1).

4.10   Meeting the UNESCO requirements

The concept of employing the HBIM to satisfy UNESCO's World Heritage nomination file requirements is based on the integration of different data sources within a common interactive environment (the integration with different data sources was described in section 2.5.7). This interactive environment can allow the users from UNESCO or other involved organisations to extract and to generate the requirements that they need directly from one database (Figure 4.91). There is also the capacity to add reports, comments, and enquiries for the participants of the nomination file.

The possible advantages of employing this method will benefit UNESCO, the involved organisations, and the participants who prepare the nomination file. For example, the key expectations for the participants involve clear time frames and process to follow. This results in reducing the cost and the time necessary for preparing the file requirements, sharing the information, as well as controlling and managing the tasks and the manpower. Further, reducing the errors and the conflicts with the tasks during the nomination file preparation are also important. Employing this method also provides easy access to the information and enables easy modification in case of any issues during preparing the file, as well as providing an avenue for any feedback from the UNESCO organisation.

The key benefits for the involved organisations can also include early follow-up of the preparation steps for the file via the HBIM environment, as well as saving time, money, and effort in sending frequent investigations to the heritage site. The key priorities for UNESCO are to reduce the cost and the time involved in preparing such files, and to reduce the time involved in the nomination process whilst increasing the credibility of their decisions.

Table 4.2 summarises how UNESCO WHNF requirements can be met within the Jeddah Heritage BIM model. The next sections will describe in detail the method of providing these requirements within the JHBIM model.

Figure 4.89 3-D JHBIM of Radwan House after the rendering (the pilot case study)

Figure 4.90 3-D JHBIM of Nasif Historical House after the rendering (the main case study)

Table 4.1

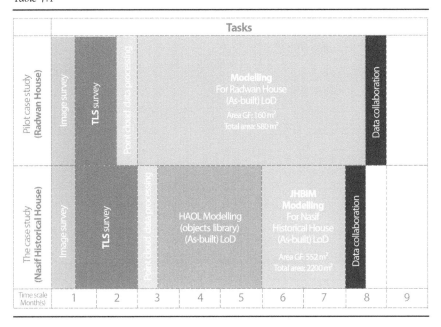

				Tasks						
Pilot case study (Radwan House)	Image survey	**TLS** survey	Point cloud data processing	**Modelling** For Radwan House (As-built) LoD Area GF: 160 m² Total area: 580 m²				Data collaboration		
The case study (Nasif Historical House)	Image survey	**TLS** survey	Point cloud data processing	HAOL Modelling (objects library) (As-built) LoD		**JHBIM Modelling** For Nasif Historical House (As-built) LoD Area GF: 552 m² Total area: 2200 m²		Data collaboration	
Time scale Month(s)	1	2	3	4	5	6	7	8	9

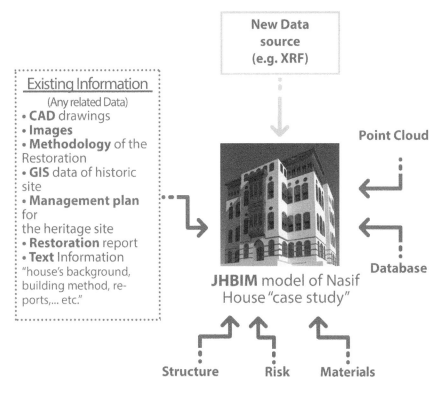

Figure 4.91 The proposed dataset for Jeddah Heritage's BIM model

Source: Author

Table 4.2

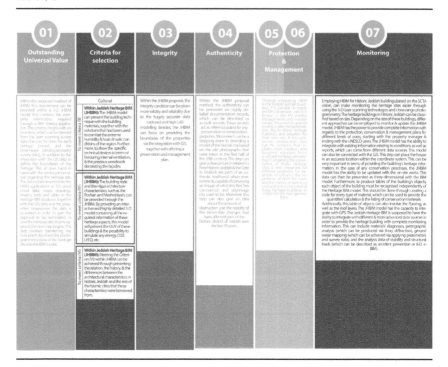

4.10.1   Level of users

Through the JHBIM model, several levels of users' access can be provided, starting with the administrators, local authorities, international regulators, and the public users (Figure 4.92). Each of these users has their own level of access, authority, and responsibility, for example, the local authorities are responsible for preparing and meeting the requirements of the nomination file. Then, these requirements are inserted into the HBIM database to be examined by the international regulators. The international regulations have access to examine, check, and report any issues with the file. The administrators have access to update, upgrade, and provide system support, as well as controlling the data. Finally, the public users can view the model and the data without any possibility of modifying or changing the data. Figure 4.93 shows the level of users' access and the responsibilities of each user group.

4.10.2   Outstanding Universal Value (OUV)

The OUV is the most important aspect and the basis of almost all the requirements of UNESCO's WHNF. Providing the OUV within the current method was based on the inventory of the property, existing information, and defining and

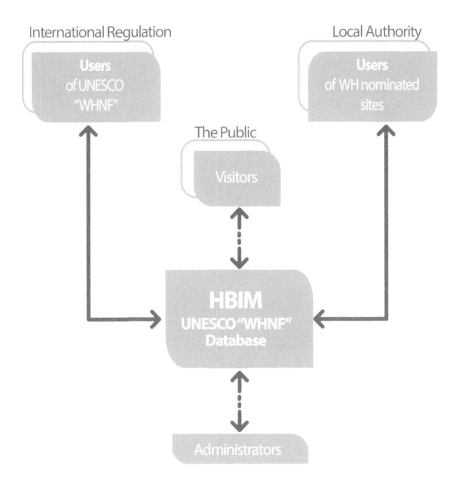

Figure 4.92 The outline of the JHBIM users
Source: Author

understanding the property (see Table 2.1: UNESCO World Heritage nomination file requirements). These requirements can be found in the form of drawings, maps, texts, images, and tables.

Regarding the case study of Historic Jeddah, the OUV was provided through the unique development of the Red Sea architectural style and the preserved urban fabric of the historic city, and via its symbolic role as a gateway to the holy city of Makkah for pilgrims reaching the Arabian Peninsula with boats over the centuries. The Roshan and Mashrabiyah have decorated most of the Historic Jeddah tower houses since the 16th century. These elements are an outstanding evolution of these coral stone houses, which are always constructed as large and complex wooden casements.

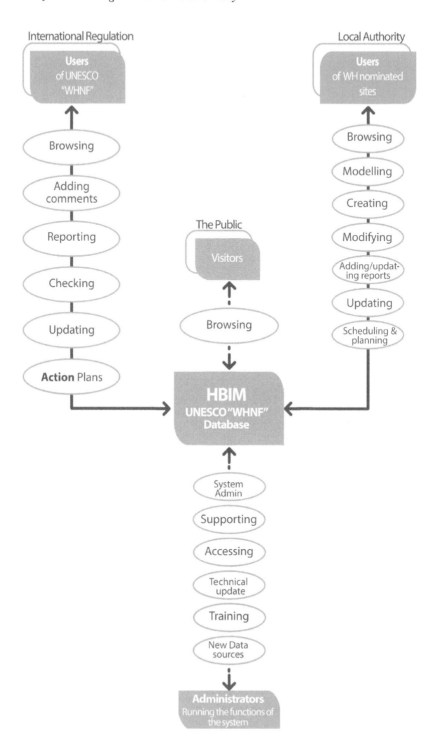

Figure 4.93 The level of users' access and the responsibility of each user group

Source: Author

The previous nomination file of Historic Jeddah dealt with this topic by providing the OUV via a GIS boundaries map, laser scanning for the main historic paths' elevations, 2-D drawings and sketches for some heritage buildings, images, and reports.

Within the proposed method of JHBIM, these requirements can be provided within a 3-D JHBIM model that contains the complete information required through a BIM sharing application. The process begins with an inventory, which can be derived from the laser scanning survey data (this can be done for each heritage house), and the close-range photogrammetry survey data (the first solution introduced using close-range photogrammetry for the inventory, see section 3.3.1), in addition to the integration with the GIS data to define the boundaries of the heritage sites (the second solution introduced using GIS for the inventory, see section 3.3.2). Figure 4.94 shows the layout of the OUV step within the proposed method of Jeddah Heritage BIM.

This all goes hand in hand with the existing information regarding the heritage site. This data is then inserted into the HBIM applications as 3-D point cloud data, maps, drawings, images, texts, and tables in the heritage BIM database and along with the GIS data and the properties' boundaries. This data is then evaluated in order to gain the approval for the site to be nominated.

Figure 4.95 shows the concept of the integration between JHBIM and 3-D GIS in Autodesk InfraWorks.

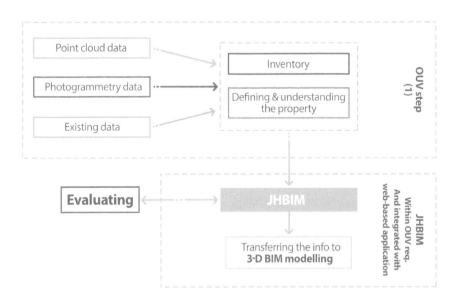

Figure 4.94 The layout of the OUV step within the HBIM method

Source: Author

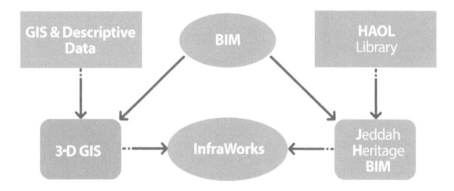

Figure 4.95 The concept of the integration between JHBIM and 3-D GIS in Autodesk InfraWorks

Source: Author

Figure 4.96 Buildings' height analysis in the selection area in Autodesk InfraWorks

Figure 4.96, Figure 4.97, and Figure 4.98 show the integration of the vector and raster data with the JHBIM of Nasif Historical House within Autodesk InfraWorks.

In case the heritage site obtains approval, the next step begins. This step involves transferring the laser point cloud and the photogrammetry data of the heritage site into the BIM model.

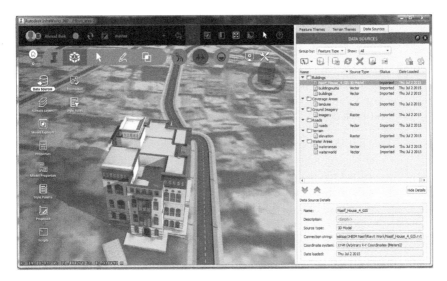

Figure 4.97 Integrating vector and raster data with the JHBIM of Nasif Historical House within Autodesk InfraWorks

Figure 4.98 The quality of the level of details in Autodesk InfraWorks

4.10.3 Criteria for selection

The second main requirement for inclusion onto the list of the World Heritage Sites (WHS) is to meet at least one out of ten criteria, which includes four natural and six cultural criteria (see Table 2.1: UNESCO World Heritage nomination file requirements). In the case of heritage buildings, the six cultural criteria seem more relevant.

Regarding the case study of Historic Jeddah, it has the criteria (II), (IV), and (VI) (see Table 2.5: WHNF requirements for Historic Jeddah nomination file). These criteria were provided as text information, maps, and images.

> *To meet Criterion (II):* "Exhibit an important interchange of human values, over a span of time or within a cultural area of the world, on developments in architecture or technology, monumental arts, town-planning or landscape design." Within Jeddah Heritage BIM (JHBIM):
>
>> The JHBIM model can present the building technique including the building materials used, together with the methods that have been utilized to combat the extreme humidity and heat climatic conditions of the region. Furthermore, the model can present the specific technical devices in terms of favouring internal ventilation, and the precious woodwork decorating the façades.
>
> *To meet Criterion (IV):* "Be an outstanding example of a type of building, architectural or technological ensemble or landscape which illustrates (a) significant stage(s) in human history." Within Jeddah Heritage BIM (JHBIM):
>
>> The building style and the Hijazi architecture characteristics, such as the Roshan and Mashrabiyah, can be provided through the JHBIM. By providing an interactive and highly detailed 3-D model containing all the required information of these heritage aspects, this model will present the OUV of these buildings and the possibility to simulate any energy, CO^2, LEED, etc.
>
> *To meet Criterion (VI):* "Be directly or tangibly associated with events or living traditions, with ideas, or with beliefs, with artistic and literary works of outstanding universal significance." Within Jeddah Heritage BIM (JHBIM):
>
>> Historic Jeddah is directly associated, both at the symbolic intangible level and at the architectural and urban level, with the Hajj, (the yearly Muslim pilgrimage to the holy city of Makkah) both at the symbolic intangible level and at the architectural and urban level. The association with the Hajj is also very evident in the urban structure of the nominated property, that includes the traditional souks running East-West from the sea to Makkah Gate, the "Ribats" and the "Wakalas" that used to host the pilgrims. The association is also evident in the architecture, notably in the façades and internal structure of the houses; and in the very social fabric of the city, where Muslims from all over the world mingled, lived, and worked together. This encouraged intangible and tangible relationships, demonstrating the intimate and long-lasting connection between the pilgrimage and the nominated property, which is a great example of why the city has such a rich cultural diversity that was caused by this unique annual event in the whole Islamic World.

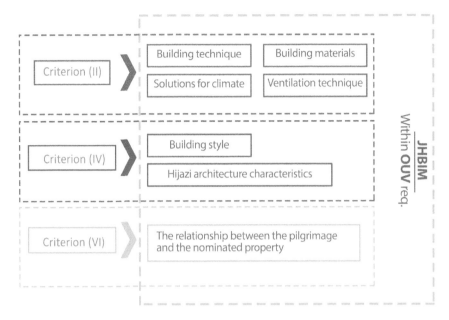

Figure 4.99 The layout of meeting the criteria within the HBIM method
Source: Author

This tangible relation seems difficult to represent within JHBIM; however, through focusing on and analysing the architectural characters of the heritage building in Jeddah, the tangible relation can be recognised through the history and the origin of where these architectural characters were borrowed from.

Therefore, meeting the Criterion (VI) within JHBIM can be achieved through presenting the relation, the history, and the differences between the architectural characteristics in Historic Jeddah and the rest of the Islamic cities from which these architectural characteristics were borrowed.

Figure 4.99 shows the layout of meeting the criteria within the Jeddah Heritage BIM method.

4.10.4 The integrity condition

The World Heritage Resource Manual describes integrity as "a measure of the completeness of the attributes that convey outstanding universal value." Thus, to ensure the integrity of the heritage site, the OUV needs to be provided.

Within the current nomination file of Historic Jeddah, the file focused on the three key concepts of integrity, which were mentioned in the UNESCO manual: the first concept being "Wholeness", the second "Intactness", and the third "Absence of threat". The nominated properties in Historic Jeddah contain several

Figure 4.100 The outline of meeting the integrity requirement
Source: Author

attributes that deliver its OUV with the requirement of "wholeness" required for the WH properties. This includes (a) examples of Jeddah's Roshan tower buildings with the ensemble of their aesthetic and functional patterns; (b) the traditional building techniques and precious woodwork decorating the façades, along with a multi-cultural population and a strong trade-based economy; and (c) the high value of the city as a link between the East and the West acting as a gateway to the holy city of Makkah. Although the heritage buildings in Jeddah are facing many issues regarding the deterioration of their historic structures, these nominated properties still possess all the required attributes to achieve the aspect of "Intactness".

Today, huge efforts are being made by the Municipality of Jeddah and the SCTA to preserving the historic district via many conservation and restoration projects of the main houses and mosques. These projects can be a favourable environment for new methods to be employed.

Within the JHBIM proposal, the integrity condition can be given more validity and reliability due to the highly accurate data captured and the high LoD modelling. Besides, the JHBIM can focus on providing the boundaries of the proprieties via the integration with GIS, together with offering a preservation and management plan. Figure 4.100 shows the outline of meeting the integrity requirement within the Jeddah Heritage BIM.

4.10.5   *The authenticity condition*

According to the World Heritage Resource Manual, authenticity presents "the link between attributes and potential outstanding universal value" and that this "needs to be truthfully expressed so that the attributes can fully convey the value of the property." The Resource Manual also describes authenticity as "a measure of how well attributes convey potential outstanding universal value" and it "can be compromised if the attributes are weak; communities cease to thrive, buildings collapse, traditions disappear, and so on". Therefore, it is very important to validate the authenticity of archaeological sites to explicitly convey their value as archaeological remains. However, many instances of reconstruction of sites often hinder the process and compromise authenticity for the proposed reconstruction. In addition, it can be difficult for incomplete reconstructions of buildings and structures to fully convey their original meaning, although this can be acceptable in some circumstances.

Regarding the Historic Jeddah nomination file, the idea of including the modern buildings, people, and commercial activities within the authenticity was presented side by side with urban centres composed of historic buildings and a traditional urban fabric, in the historic district of Jeddah. Additionally, the authenticity of heritage buildings in Jeddah is particularly evident as most are relatively well preserved. Despite the fact that over time these heritage buildings have been faced with a number of issues and challenges, these buildings have not been significantly altered with modern additions and in-depth transformations that could affect their shape as well as their substance.

Within the JHBIM proposal method, the authenticity is able to be presented via highly detailed documentation records, which can be described as as-built records. These records act as reference points for any preservation or conservation purposes. Moreover, it can be the next step for providing a model of the historic city based on the old photographs that were taken in the first half of the 20th century. This step can give a clear picture in relation to how Historic Jeddah and the Gate to Makkah are parts of an authentic traditional urban environment, capable of conveying an image of what this Red Sea commercial and pilgrimage city used to be. Additionally, this step can also give an idea about the amount of destruction and the rapidity of the irreversible changes that have affected parts of the historic district of Jeddah over the last 50 years.

Figure 4.101 shows the outline of meeting the authenticity requirement within the Jeddah Heritage BIM.

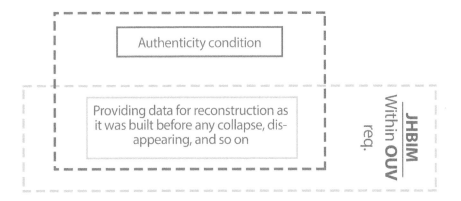

Figure 4.101 The outline of meeting the authenticity requirement
Source: Author

4.10.6   *Building protection and management*

The fifth main requirement is ensuring effective legal and traditional protection, so that the property has the best available protection within a clear jurisdiction and context, but layers of legislation and other protection can sometimes be necessary. It is very important that the legal and traditional protection work together effectively to produce a layered method for protection, as well as legal protection, which can offer a suitable and supported context to safeguard the traditional features when sites face any threats (for more, see Table 2.1: UNESCO World Heritage nomination file requirements). The sixth main requirement is to focus on the management of the World Heritage property. This includes both the features and attributes that are related to achieving an OUV of the nominated site. According to the Resource Manual, the object is "to ensure that the value, authenticity and integrity of the property are sustained for the future through managing the attributes," so it is very important to deliver management plans of the potential OUV of the site that are fully coordinated and linked to the preservation requirements of the site (for more, see Table 2.1: UNESCO World Heritage nomination file requirements).

Regarding the case study of Historic Jeddah, the Saudi Commission for Tourism and Antiquities (SCTA), and in coordination with the Jeddah Municipality, as well as the participation of the civil society, have drawn up a general strategy for the preservation and revitalisation of the area. The local branches of the Jeddah Municipality and the SCTA are responsible for the daily management of the nominated property, including organising the protection, maintenance, cleaning, and presentation of the historic site (for more, see Table 2.5: WHNF requirements for Historic Jeddah nomination file).

Through employing HBIM to the Historic Jeddah buildings, it is possible to Apply the 6th Dimension, which is known as the FM (Facility Management) and maintenance (section 2.4.3.2 explains more about FM). This could be applied during or after the preservation step for these heritage buildings and adapted for their new use. A fully preventive preservation plan can be enforced that includes short-, medium-, and long-term sub-plans.

From a cultural heritage perspective, maintaining the preservation of these heritage sites is very important for different purposes; for example, to enhance the experience of both the users and the visitors, as well as to save and restore the value of these heritage buildings in Jeddah for the next generations.

4.10.7   *Building monitoring*

The seventh main requirement is having good management for monitoring, as (Marshall, 2011) says, this includes "a range of key factors, which will give an indication about the current situation of the property, its state of conservation and its likely future". The Resource Manual also states that monitoring "provides valuable information for the property manager, which can show that protection, conservation and management are achieving their goals or that changes need to be made". Therefore, monitoring should focus on the protection, management, authenticity, and integrity for heritage site properties. Every six years, the World Heritage Committee requires a monitoring report, so that key indexes should be included in the nomination file in order to evaluate and measure a range of aspects, also linked with the state of preservation in reference to the nominated property.

Each of these indexes must be linked to the properties that deliver potential OUV to confirm that they are being preserved, managed, and protected to sustain this OUV. Based on a clear time frame, monitoring should be undertaken regularly and be suitable for the character of the nominated site. The monitoring results could be influenced by the reliability or apparent reliability of those who undertake the monitoring, so that if the monitoring report is completed with transparency by independent experts, the monitoring results will have greater credibility (Marshall, 2011) (for more, see Table 2.1: UNESCO World Heritage nomination file requirements).

Regarding the case study of Historic Jeddah, according to UNESCO, "Monitoring is an activity aimed at regularly assessing the condition of the site and progress made, or difficulties encountered, to implement the activities proposed via standard scientific protocols". In the case of Historic Jeddah, the Gate to Makkah, and based on the SCTA vision, the monitoring "allows the record of changes at two scales: the larger cityscape in which the site is located and the actual management of the nominated property". Furthermore, by linking the results of monitoring to management decisions, the urban authorities can find the best way to use funds and staff in the nominated property.

For an urban site such as Historic Jeddah, the Gate to Makkah, according to SCTA (2013), "three distinct and complementary sets of indicators, ranging from conservation, to social statistics and planning data, to verify the overall impact of the revitalization and conservation strategy proposed for Historic Jeddah, the Gate to Makkah: urban and architectural conservation indicators, social indicators, planning and development indicators, urban and architectural conservation indicators, and the record of environmental data, offers essential information to be crossed with site deterioration" (for more, see Table 2.5: WHNF requirements for Historic Jeddah nomination file).

Within employing HBIM for Historic Jeddah buildings, and based on the SCTA vision, monitoring the heritage sites can be made easier through using the 3-D laser scanning technologies and close-range photogrammetry. Furthermore, the heritage buildings in Historic Jeddah can be classified based on size, and depending on the size of these buildings, different approaches can be employed to monitor and to update the JHBIM model.

Jeddah Heritage BIM has the potential to provide complete information with regards to the protection, conservation, and management plans for different levels of users, starting with the property manager and ending with the UNESCO users, in order to achieve their aims and/or facilitate any changes that are needed to be made. Additionally, the JHBIM model has the ability to integrate with existing information relating to conditions, as well as reports, which can come from different data sources. This model can also be connected with the GIS. This step can provide the house with an accurate location within the coordinate system. This can be very important in terms of providing the building's heritage information. Moreover, in the case of any conservation processes, the JHBIM model has the ability to be updated with the on-site works. This data can then be presented as three-dimensional with the BIM model.

Furthermore, to produce tables of the building's objects, as Dore and Murphy (2012, pp. 369–376) described, each object of the building must be recognised independently within the Heritage BIM model. This should be done through creating a code for every type of material, which can be used to provide the quantities' calculation and the listing of conservation materials. Additionally, this table of objects can also involve the flooring, as well as the roof layers.

Along with the ability to integrate with the GIS, the JHBIM model has the capacity to integrate with the Global Positioning System (GPS). This feature can be provided through the advanced technology that has been used in mobile and tablet devices, which are compatible with the BIM application, such as BIM 360 field and Autodesk A360.

In order to support the monitoring purpose within JHBIM, another data source for collecting and analysing can be employed via the remote sensing method, which is generally categorised within aerial and photographic techniques. The first technique can be used to locate the heritage buildings with the ability of producing linear drawings manually. The second technique is dependent on the satellite sensors in order to generate a 3-D model, which is created once the geometrical errors have been received (caused by the earth's spherical shape). This technique is ideal to be employed for huge areas (such as Historic Jeddah), as it is capable of managing huge volumes of information and can provide monitoring for historic districts' buffer zones (Andrews et al., 2013).

The Jeddah Heritage BIM must have the ability to integrate with different and more advanced data sources in order to provide the heritage building with complete monitoring information. This can include materials' diagnoses, petrographic analysis which can be produced via X-ray diffraction, ground water mapping which can be achieved via applying piezometers and survey data, and the analysis data of stability and structural loads (which can be described as accident prevention or 8-D in BIM).

Figure 4.102 and Figure 4.103 show the use of the JHBIM with an iPad: interactive visualisation and queries can be easily carried out.

Figure 4.102 The use of the JHBIM with an iPad: interactive visualisation and queries can be easily carried out

Figure 4.103 The use of the JHBIM with an iPad: to visualise the 3-D point cloud data

4.11   Evaluation and testing

The evaluation and testing step was done with regard to two aspects. The first aspect included confirming the main requirements of a nomination application, using the matrix (see Table 2.1: UNESCO World Heritage nomination file requirements) to establish if the application of HBIM can assist the UNESCO procedures for World Heritage nomination file. This test was carried out interviewing the heritage practitioners of Historic Jeddah and based on documentation produced from the JHBIM. The second aspect, the accuracy testing, was carried out by comparing 2-D elevations sections and plans between the point cloud and those produced from BIM and identify errors.

4.11.1   *Heritage experts' evaluation*

This test was performed through an interview with two of the heritage experts in the historic district by using the documentation that was provided from the Jeddah Heritage BIM. This documentation included conservation drawings automatically produced from the laser scan and image survey, and HBIM. In details, these documents included elevations, plans, sections, working drawing details, and 3-D models (see section 4.8). Through interviewing two of the heritage experts the outcome engineering documents from the JHBIM model were ranked in terms of meeting UNESCO criteria.

The first expert was Engineer Sami Nawwar, the Director of Historic Jeddah Preservation Department, Municipality of Jeddah. The second expert was Engineer Sultan Faden, the Director of Historic Jeddah Office, Saudi Commission for Tourism and National Heritage (SCTH). Both experts participated in preparing the UNESCO World Heritage nomination file of Historic Jeddah (the Gate to Makkah), which was submitted to UNESCO in 2012.

Through interviewing both experts it was found that they have some knowledge regarding BIM, however, they did not know much about Heritage BIM. Moreover, by presenting the concept of Heritage BIM and showing how it can be used in the case of Historic Jeddah building, it was discovered that both experts had different opinions. For example, Engineer Nawwar was very excited to employ this technique to the heritage building in Historic Jeddah and to examine the outcome information, while Engineer Faden was more cautious about the idea and the result. Moreover, some of Engineer Faden's cautions involved the process to produce the engineering information from the HBIM model (is it automated), and how it will be operated.

These different opinions might have derived from their different experiences. For example, the first expert (Engineer Nawwar) deals with the on-site issues (see Table 4.3), while the second expert (Engineer Faden) handles the office work issues (see Table 4.4). Therefore, these opinions reflected their selections on the evaluation application of HBIM can assist the UNESCO procedures for World Heritage Nomination File.

Table 4.3

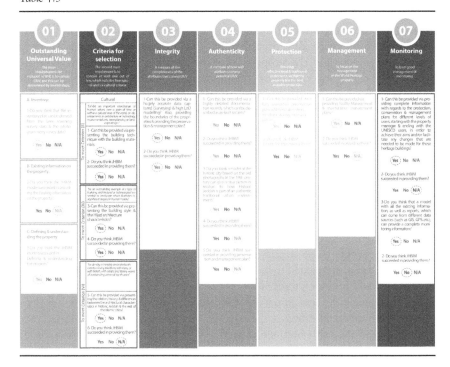

Table 4.4

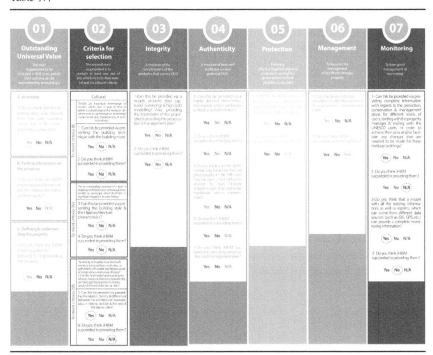

Regarding using JHBIM to provide the OUV, both experts agree that through the proposed model the "Inventory" can be provided. However, they were cautious about the integration of the existing information into the JHBIM mode. Moreover, both experts agree that through the proposed model it was easier to define and understand the property. Regarding using JHBIM to provide the second, the third, and the fourth requirements, the experts responded differently. As it has been mentioned in the nomination file of Historic Jeddah, the Gate to Makkah, the second requirement which is meeting the criteria for selection, the city has the criteria (II), (IV), and (VI). Engineer Nawwar believed that JHBIM can provide these criteria. Besides, he believed that the third (Integrity) and the fourth (Authenticity) requirements can be provided via highly detailed documentation records that were presented through the JHBIM technique. However, Engineer Faden believed that these requirements are not enough to be explained using engineering drawings, and that these drawings could just provide additional support. Regarding using JHBIM to provide the fifth, the sixth, and the seventh requirements, both experts agree that JHBIM model can provide these requirements, however, they agreed that more engineering information needed to be produced from the model. For example, these drawings would include preservation plans, Facility Management and maintenance management plans, and a database which includes the existing information of the heritage building.

By interviewing both heritage experts in Historic Jeddah, the two opinions have helped to establish a clear picture of the use of JHBIM, in order to meet the UNESCO WHNF requirements. The missing engineering information that the experts required was not presented yet, due to the time and manpower limitations. Finally, both experts came to an agreement that Heritage BIM will have great applicability for supporting and meeting the requirements of the UNESCO World Heritage nomination file.

4.11.2   The accuracy testing

The second aspect, the accuracy testing, was carried out by comparing 2-D elevations and 2-D plans between the point cloud and those produced from BIM in order to identify differences. For this test ten distances on the north elevation and on the first floor plan were chosen from the point cloud model to be compared with the JHBIM model of Nasif Historical House.

Figure 4.104 shows the selected points on the north façade of the BIM model for the accuracy comparison process. Figure 4.105 shows the selected points on the first floor plan of the BIM model for the accuracy comparison process.

Figure 4.106 shows examples of the selected points on the north façade of the point cloud for the accuracy comparison process. Figure 4.107 shows examples of the selected points on the first floor plan of the point cloud for the accuracy comparison process.

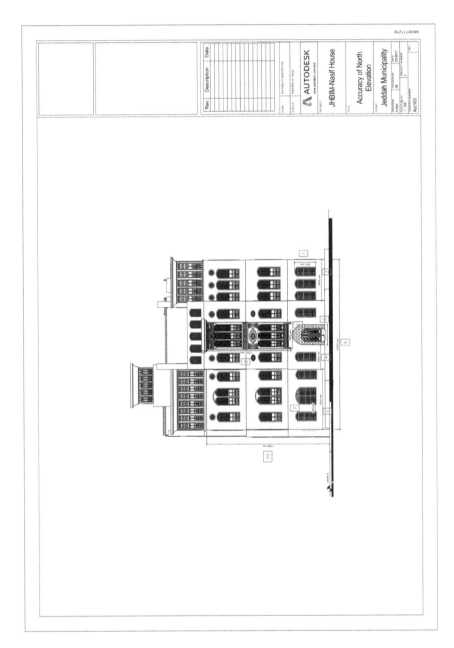

Figure 4.104 The selected points on the BIM model façade for the accuracy comparison process

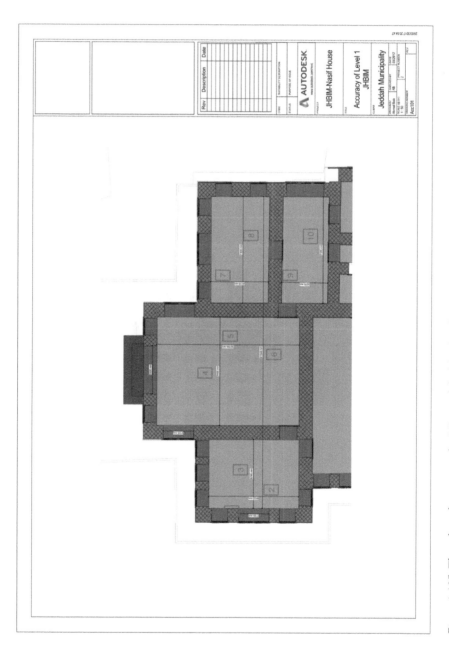

Figure 4.105 The selected points on the BIM model of the first floor plan for the accuracy comparison process

Figure 4.106 Examples of the selected points on the point cloud façade for the accuracy comparison process

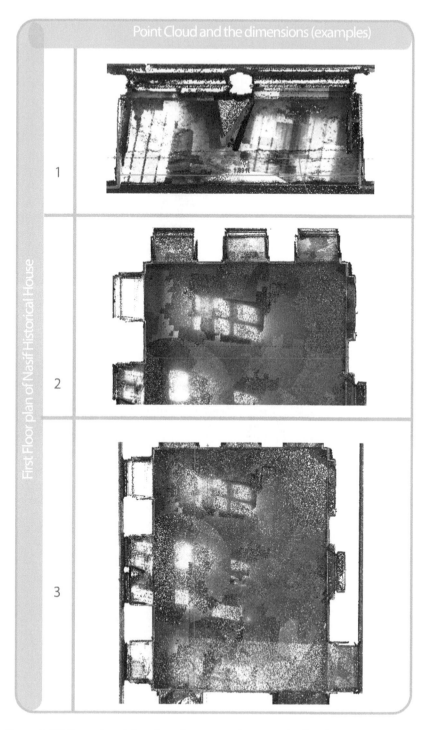

Figure 4.107 Examples of the selected points on the point cloud of the first floor plan for the accuracy comparison process

Table 4.5 Comparison of the measurements between both the point cloud model and the JHBIM model of Nasif Historical House

	Dimension No.	JHBIM Model (mm)	PC Model (mm)	Diferences (mm)	Average (mm)	Standard
First Floor Plan	1	2165	1891	274	2028	σ: 87.29
	2	6432	6498	66	6465	
	3	4746	4906	160	4826	
	4	7495	7554	59	7524.5	
	5	10184	10030	154	10107	
	6	21544	21684	140	21614	
	7	4118	4406	288	4262	
	8	7358	7622	264	7490	
	9	3278	3355	77	3316.5	
	10	7358	7473	115	7415.5	
North Elevation	1	3071	2934	137	3002.5	σ: 55.87
	2	5679	5641	38	5660	
	3	2773	2701	72	2737	
	4	3403	3390	13	3396.5	
	5	8332	8329	3	8330.5	
	6	23533	23590	57	23561.5	
	7	1053	1034	19	1043.5	
	8	3629	3661	32	3645	
	9	12781	12882	101	12831.5	
	10	17680	17850	170	17765	

Note: The numbers in italic indicate the larger measurement.

Table 4.5 compares the measurements between both the point cloud model and the JHBIM model of Nasif Historical House. As it can be noted that there are differences, some are bigger, and some are smaller. The highest difference on the measurements between two points in the BIM model and those in the point cloud model can be found in the first floor plan (distance number 1 with 274 mm for an average of 2028 mm). This high difference occurred due to an issue during creating the object in the HAOL library. The average of the differences on the measurements for the first floor plan is 159.7 mm, while the average of the differences on the measurements for the north elevation is 64.2 mm.

Furthermore, the Standard Deviation for the differences on the measurements for the first floor plan is 87.29 mm, while the Standard Deviation for the differences on the measurements for the north elevation is 55.87 mm.

4.12   Summary

The major issue that faces the historical district of Jeddah today, as pointed out by the director of the historic area, Engineer Sami Nawwar, is how the local authorities protect such buildings from the risk of collapse and erosion through ageing and human factors, as well as disasters such as fires and flood. In 2012, after the Historic Jeddah was added to the UNESCO World Heritage Sites' list, the cultural value of the historic district has increased, along with the government attention it receives. Moreover, to remain in UNESCO's World Heritage List, the heritage site must be successful in field inspections, as well as providing all the protection, management, and monitoring plans which need to be submitted with the nomination file. These requirements can be provided through employing Jeddah Heritage BIM. In detail, the concept of employing the BIM to satisfy UNESCO's World Heritage nomination file requirements is based on the integration of different data sources within a common interactive environment. This interactive environment can allow the users from UNESCO or other involved organisations to extract and generate the information that they need directly from one database. There is also the ability to add reports, comments, and enquiries by the participants of the nomination file.

The JHBIM focused on Nasif Historical House, which is considered to be one of the most important historical houses in Historic Jeddah. The project examined a number of methods for HBIM, then looked at meeting UNESCO's WHNF requirements. The first step was to examine the on-site data collection methods based on the laser scanning survey and the image survey methods, which were initially employed in S. Fai et al. (2011a, 2011b) and Murphy (2012). The second step was to examine the "Scan to JHBIM". During this step, the Autodesk Revit and Revit Family was utilised in order to build the 3-D JHBIM model, depending on the 3-D point cloud, which was different from Murphy's (2012) method and similar to Stephen Fai et al.'s (2011a, 2011b) method; however, the main difference can be found in the case study relating to the architectural style of the buildings. The main modelling approach was the manual approach for difficult and non-geometrical shapes, while the semi-automated approach was used for the sample and geometric shapes, such as some of the walls and the floors. During the scan to the JHBIM stage, a parametric objects library was developed for the Hijazi objects, which is described as HAOL. The third step was to satisfy the requirements of UNESCO's World Heritage nomination file via the use of the JHBIM model.

5 The conclusion

5.1 Introduction

Worldwide systems designed to support heritage preservation, such as UNESCO's World Heritage nomination system, have achieved significant recognition. However, these systems face a number of challenges and issues and have to continually adjust policies and strategies in order to respond to these challenges. The application for a heritage site to be listed requires the preparation of UNESCO's World Heritage nomination file. This represents a challenge for both the nominee and UNESCO because the file must contain a very detailed and thorough description of the site. For UNESCO, the large numbers of heritage sites around the world result in a significant volume of information that must be scrutinised and evaluated before the World Heritage status of a site can be acknowledged. This book has discussed the requirements (see section 2.2) and the issues (see section 3.2) regarding the preparation of a UNESCO World Heritage nomination file (WHNF), and why it is important to have new methods for preparing the nomination file (see section 3.4). In additon, the book discussed how Heritage BIM (HBIM) can solve the issues linked with providing all the required information in relation to these sites that will be listed in the UNESCO World Heritage Sites List (WHSL) (see section 3.6).

5.2 Heritage BIM as a new model for the World Heritage nomination file

Meeting UNESCO's World Heritage nomination file requirements leads to many challenges in relation to missing information, which can be linked to the heritage site, the parties involved, the UNESCO procedure, and technical and economic issues (see section 3.2). Over the last few decades, several solutions have been offered to overcome these problems and to satisfy some of the UNESCO World Heritage requirements (see section 3.3). These solutions have been focused on issues such as the criteria, OUV, states parties, cultural governance as a method for social and cultural sciences, the politics of heritage, and the decision-making process (Dumper and Larkin, 2012; Jokilehto, 2015; Meskell et al., 2015; Schmitt, 2009). However, a limited number of solutions have been proposed with regards to technological and informatics solutions and how to develop a mechanical method to meet the requirements. These solutions include those provided by D'Andrea et al. (2012), El Garouani and Alobeid (2013), Gillot and Del (2011), and Santana-Quintero and Van Balen (2009). Through focusing on these solutions, their strengths and weaknesses can be noted. For example, the proposed solution for a pre-inventory plan via employing the photogrammetry method which was proposed by Santana-Quintero and Van Balen (2009) has a number of strengths and weaknesses. The strengths are methods that lead to the speeding up of the definition, identification, and dissemination of the basic materials of the historic sites. On the other hand, the weaknesses are methods that cannot be used for advanced proposes and analyses such as quantity estimating, or the structural loads. In contrast, the second proposed solution for inventory plans via GIS which was proposed by Laurence Gillot and Andre Del (2011) is strong on the analyses side, but weak in the 3-D visualisation side. The third proposed solution is dealing with the site complexity via DSM, proposed by El Garouani and Alobeid (2013), which focuses on the environmental and urban planning sides more than the architectural characteristics of the heritage site. The fourth proposed solution for 3D-ICONS, which was proposed by D'Andrea et al. (2012), is strong in providing a complete 3-D content model and providing remote access for the public to the complex models. However, this proposed solution is weak on the ACE engineering sides, and it seems this method focuses more on the visualisation aspect.

One observation is that if these solutions were to be combined together into a common method, it could be more effective. For example, this advanced method can be defined as a complete 3-D content model, such as in the fourth proposal, including the ACE engineering information. With the power of the GIS and DSM, as presented in the second and third proposals, one can do all the simulation and the analyses can be finalised, with the ability to provide a complete pre-inventory plan of the heritage site in a short time, such as in the first proposal.

Thus, it is very important for a common new approach to be developed in order to meet UNESCO's WHNF requirements.

The research question in this book seeks to propose a combined new methodology. *How can Heritage Building Information Modelling (HBIM) provide innovation in creating missing information for UNESCO's World Heritage status? And what additional cultural value can a sustainable update of HBIM provide for such sites?*

The theory of applying HBIM to satisfy the requirements of UNESCO's World Heritage nomination file was based on developing an interactive approach to move from traditional 2-D constructive representation and 3-D content models to HBIM in order to support preventative conservation, information sharing, and knowledge dissemination in relation to heritage, for professional users, institutions, and experts involved in UNESCO's World Heritage nomination file process (Eastman et al., 2011; Fussel et al., 2009). Furthermore, Heritage BIM has the capacity to integrate with different heritage buildings (in BIM format) in order to provide comprehensive information in relation to the heritage buildings in the historic district. HBIM will offer an integrated approach to the overall design, construction, and post-construction process in the architectural and engineering industries (Eastman et al., 2011; Saygi et al., 2013). In addition, HBIM will automatically provide complete engineering information and drawings. According to Henry Owen-John, the Head of International Advice for English Heritage, the existence of a BIM system that helps understanding of a historic building and how best it should be managed and protected, will not, of itself, make the case for World Heritage Site status. It is more a case of identifying a historic structure or structures which are of sufficient importance to merit consideration for inclusion on the World Heritage List. On the other hand, it would appear that there are significant challenges facing those applying BIM in the historical field, including organisation, technical obstacles, and site issues. Another important point regarding applying BIM into the historic field is determining suitable levels of detail (LoD).

In the case of historical buildings, one of the most significant points is to rebuild the past, which requires a very high level of detail (i.e. As-Built level). This level of detail can be achieved based on advanced technologies, such as laser scanning and photogrammetry. These technologies can provide a very rich cloud point model, which can be used as a basis for the BIM model.

Next, we will focus on the first sub-question. *For which aspects of the UNESCO nomination file can HBIM provide highly accurate fully documented information at the scale required for the UNESCO nomination project?*

By employing BIM in the heritage sector fully documented information can be provided. This has been achieved for a number of heritage sites around the world, such as in the projects of Barazzetti et al. (2015), S. Fai et al. (2011a, 2011b), Macher et al. (2014), Ma et al. (2015), Murphy (2012), Oreni (2013), Oreni et al. (2014), and Penttilä et al. (2007) (see section 3.5). Furthermore, by employing the HBIM for a heritage building, the lifecycle of the heritage building can be extended through being able to offer maintenance plans, protection and management plans, as well as follow-up access (Fregonese et al., 2015) (section 3.8 explains more about the heritage building lifecycle). Moreover, by focusing on the concept of BIM (in section 2.4) and HBIM (in section 3.5), the advantages of using HBIM to meet UNESCO's WHNF requirements become clear. They are:

1 Within the HBIM solution, inventory plans can be directly derived from the 3-D survey data from the laser scanning (Cheng and Jin, 2006) (see section 3.6.1).
2 The provision of protection and management plans (see section 3.6.2).
3 The provision of conservation documentation automatically after completing the HBIM models (see section 3.6.3).
4 Offering a standardisation form in order to prepare the nomination files. Through inputting the information into the BIM process step by step as is required (the UNESCO WHNF requirements are mentioned in section 2.2), a common structure form containing the complete information that is needed can be generated (see section 3.6.4).

5   The provision of building monitoring in different stages of the building life-cycle. Through the BIM field management applications, heritage buildings can be monitored (see section 3.6.5).

6   The capability to integrate different data types (see section 3.6.6).

7   The provision of maintenance and refurbishment plans for the historic building through BIM facility management (FM) (Ilter and Ergen, 2015) (see section 3.6.7).

8   The provision of knowledge regarding the development and understanding of how the building has evolved and what is significant about it in historic and architectural terms (see section 3.6.8).

9   The provision of access to the sharing and updating of data with UNESCO and stakeholders in an attempt to prepare the best and most suitable World Heritage nomination file. Moreover, the integration of different sources of data and accessing them within the HBIM frame and application can offer new ways of updating the WHNF (see section 3.6.9).

10   The capability to make the reconstruction of a 3-D building and object become reality (great visualisation) (see section 3.6.10).

11   The provision of full engineering information and drawings (see section 3.6.11).

12   Providing a tool for facility management (FM) of the heritage building. This can be achieved via supporting preventative conservation and providing the maintenance operations (Cheng et al., 2015) (see section 3.6.12).

5.3   Working toward JHBIM

The second and the third sub-questions were: *How can HBIM be used to manage and monitor historical buildings?* and *How can HBIM be used to better maintain, protect, and record the updated information of the historical buildings?* Regarding providing a new model for the UNESCO World Heritage nomination file, Historic Jeddah, Saudi Arabia's file was selected as the case study for this book, which is referred as Jeddah Heritage Building Information Modelling (JHBIM) focusing on the Nasif Historical House. The main reason for choosing this case study is due to the big gap in prior knowledge in relation to the Hijazi heritage buildings in the Middle East.

Many of these buildings have no engineering data in order for them to be restored in the case of collapse or any other disasters. The JHBIM model was based on on-site survey data, which included the laser scanning survey and image survey, which have been employed in other studies (S. Fai et al., 2011a, 2011b; Murphy, 2012), as well as the architectural Hijazi pattern books and any useful existing data. The output data from the on-site survey data was subsequently evaluated by the administrator. This step is similar to the justifying of the OUV step.

During the modelling process, the Hijazi Architectural Objects Library (HAOL) was developed. The HAOL was based on the Revit family. Within the HAOL library, the architectural rules were developed and tested based on the

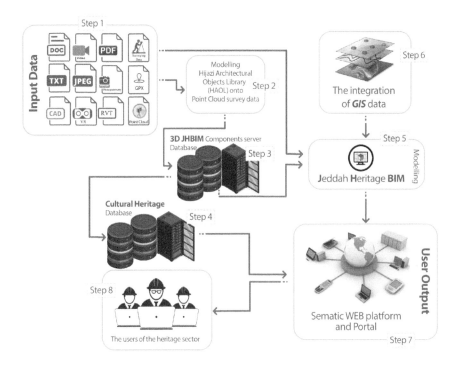

Figure 5.1 The outline of designing the systems architecture for GIS WEB dissemination

Hijazi pattern, alongside the laser scanning and image survey. The objects in the library were as highly detailed as "as-built level of details". The objects in this library can therefore be inserted into any project in the historic area of Jeddah in the future.

By the end of the JHBIM process, the non-geometric data and the existing information, such as restoration reports, GIS data, images and text data, integrated with the JHBIM model in order to satisfy the requirements of UNESCO's World Heritage nomination file and to provide complete engineering drawings and information to the Jeddah Municipality.

Figure 5.1 shows the outline of designing the systems architecture for GIS WEB dissemination so that the required procedures/criteria for the UNESCO process can be contained and updated in the WEB/GIS JHBIM. Furthermore, this information can be used for all conservation and preservation purposes in the future.

5.3.1 JHBIM versions

As has been mentioned throughout this book, the Jeddah Heritage BIM is an interactive model. Therefore, this model can be remotely managed and reviewed.

In order to complement this concept, the model can be provided in a number of versions. These will be detailed next.

- *BIM version:* this was the basis for the model, which contains all the architectural, structural, and infrastructural information in a complete 3-D model (Figure 5.2).
- *A360 version:* after modelling the 3-D model in the BIM version, the 3-D model was inserted into Autodesk A360. This enabled the collaboration between the various parties and organisations involved. The A360 version allows those involved to view, comment, share, mark-up, review, and find all the heritage site information in one place (Autodesk, 2016b).

   Figure 5.3 and Figure 5.4 show screen shots of the A360 application on the iPad, illustrating some of the features of Autodesk A360 BIM®.
- *Autodesk 360 Glue version:* 360 Glue can be described as a cloud-based BIM application for management and collaboration, which links all project members, as well as streamlines the BIM project workflows, starting with pre-construction up to the phase of construction execution (Autodesk, 2013a). The 3-D Revit model was linked to the Autodesk BIM 360 Glue on iPad. This provides a model of multi-disciplinary coordination and allows for the detection of clashes, as well as offering access for stakeholders across the scheme's

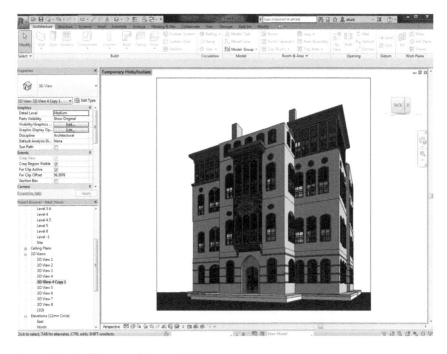

Figure 5.2 Nasif Historical House BIM version

Figure 5.3 Showing the properties of the main Roshan within Autodesk A360 using the iPad app

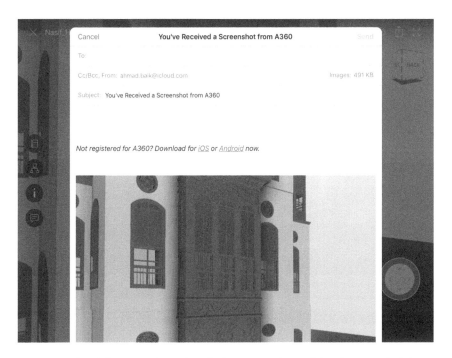

Figure 5.4 The ability to send emails with the issue in the model within Autodesk A360 on the iPad app

lifecycle. On a global scale, architects, engineers, owners, and developers can collaborate in real time from the office or via mobile devices.

Moreover, they can upload models from Autodesk Navisworks, Revit, AutoCAD and Civil 3-D, and literally "glue" them together in the cloud (Autodesk, 2013a). At this stage, Autodesk BIM 360 Glue was used to present and explore the 3-D model's details, to enhance understanding, and to mark up models and verify the dimensions online or offline for the heritage building in Historic Jeddah. The Autodesk BIM 360 Glue features include:

1   Reviewing complex multi-disciplinary project models.
2   Help with identifying and avoiding potential problems.
3   Connecting workflows with Autodesk Navisworks Manage.
4   Helping to isolate clashes within certain Autodesk products.
5   Reviewing models, asking questions, and sending notifications.
6   Intuitively exploring project model details.
7   Marking up models and verifying dimensions online or offline.
8   Linking documents directly to intelligent BIM objects.

Figure 5.5, Figure 5.6, Figure 5.7, and Figure 5.8 show screen shots of the BIM 360 Glue mobile application on the iPad, illustrating some of the features of Autodesk BIM 360 Glue.

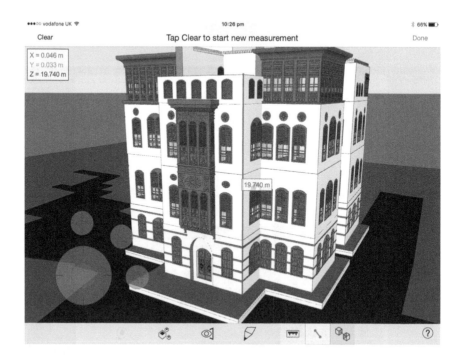

Figure 5.5 BIM 360 GLUE® shows the Diminution features

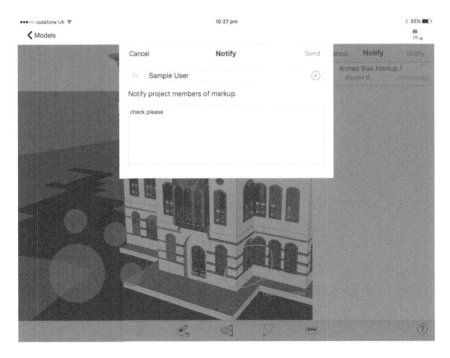

Figure 5.6 Reporting an issue within BIM 360 Glue

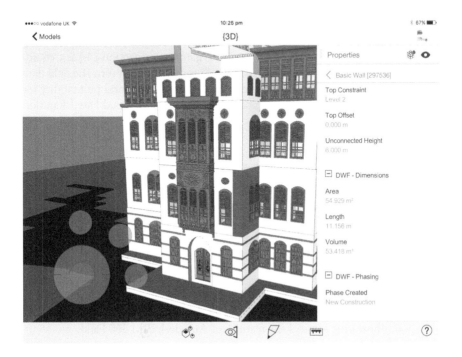

Figure 5.7 BIM 360 GLUE® shows some features

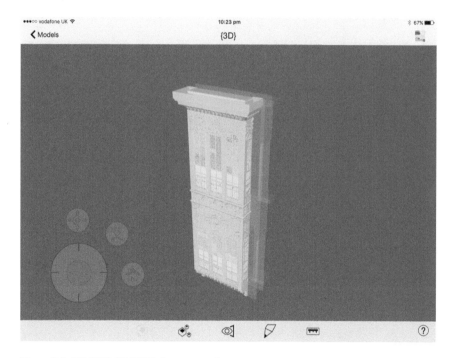

Figure 5.8 BIM 360 GLUE® shows some features

- *Autodesk 360 Field version:* the iPad application of Autodesk BIM 360 Field is a field mobility tool that is aimed at allowing field-level access to information and to enable collaboration with regards to inspections, issues, equipment, and tasks to be completed. The Autodesk app can turn the field data into information that improves quality, security, safety, and profitability for construction and projects' capital: depending on the cloud-based function (Autodesk, 2013b).

  In this stage, the BIM 360 Field was used to present the data online, to mark up issues with images taken directly from the field, and also, to manage the information. The Autodesk BIM 360 Field features include:

1 Field data management.
2 Visuals that accelerate field communication.
3 BIM updates based on field data.
4 An interactive project website.
5 Quality and safety management.
6 Commissioning.
7 Pushpins for tracking issues in 2-D and 3-D.

Figure 5.9 shows a screen shot of the iPad Autodesk BIM 360 Field application, illustrating some of the features of Autodesk BIM 360 Field.

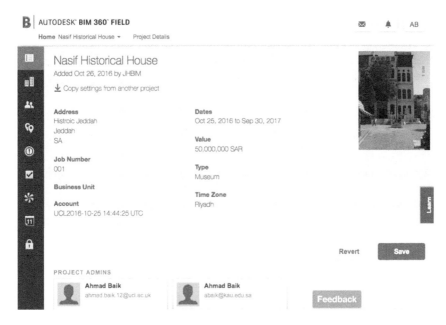

Figure 5.9 Project setup within Autodesk BIM 360 Field

5.4 Improving the cultural value of heritage buildings

Employing BIM for heritage buildings (HBIM) can provide the answer for the fourth sub-question, which is, *How can HBIM improve the cultural value of heritage buildings in the short, medium, and long term, as well as provide a better future for historical buildings?*

As has been discussed, HBIM can achieve these objectives through providing the heritage buildings with a lifecycle approach to support different aspects during the timeline of these heritage sites. These aspects can include facility operation and Facility Management (FM), which can support the construction management, project management, and cost management for any reconstruction or refurbishment in relation to the heritage buildings.

In addition, lifecycle management can be used to examine the future development plans within the context of these heritage buildings. Moreover, the HBIM model is created with all the available data regarding the building within different periods "in the past and present", which can be described as a digital archive, and through the HBIM model, different operations can be undertaken in order to provide a better future for the heritage building (Stephen Fai et al., 2011a).

With the purpose of providing a lifecycle approach for heritage building within the BIM environment, a Heritage Building Lifecycle Management (HBLM) system can be put into place. The HBLM sets into practice the level 3 BIM method, which provides a highly efficient extended collaboration model referring to Heritage Building Lifecycle Management (HBLM) and the heritage manufacturing industry best practice (Moriwaki, 2014).

Through providing the heritage buildings with a lifecycle and showing how these buildings can be managed in terms of different aspects such as conservation, operations, and rehabilitation for better uses, this leads to an increase in the value of the heritage buildings.

5.5   The impact of HBIM on Digital Cultural Heritage

The fifth sub-question is, *How does HBIM impact on Digital Heritage?* As discussed during the last decade, technological developments have had a huge influence on many aspects of life, such as communication, transportation, and education, and this influence has also reached new areas such as the architecture industry and the heritage field. Through the enormous developments in architectural and survey technology, these changes have been reflected in their expanding usage, which includes use in the heritage field. Moreover, during the last few years, the use of digital photos for close-range photogrammetry has been employed for several historical sites around the world to capture the details of historical objects. In addition, 3-D laser scanning has been introduced to the heritage sector in order to scan and create accurate as-built 3-D point cloud models. In this case, according to Garagnani and Manferdini (2013), these 3-D models that are created by the 3-D laser scanning method are suitable for the purposes of periodic checks with regards to the structure's conservation, and most importantly, for documentation purposes.

Recently, huge improvements in BIM applications have allowed for the integration and conversion of laser scanning data within these modelling software such as Autodesk Revit and ArchiCAD. This development provides huge opportunities for employing BIM in the heritage field and further, to the existing building sector. However, between the 2000s and 2010, the research on this topic (using BIM for existing buildings and in the heritage field) was very limited and only focused on the countries that were already adopting BIM in their systems. On the other hand, after 2011 until the present day, a huge number of projects and studies have been published in a relatively short time. The interesting point is that a number of these HBIM studies and projects have been undertaken in countries that do not currently support BIM in their systems.

By focusing on the main purposes of these HBIM studies and projects, it can be noted that the common goal is to provide digital documentation for the heritage site. This information can then be used for different purposes at the current time or for future analysis. In addition, based on the main objective of Digital Architectural Heritage (DAH), which aims to digitalise heritage for the next generations, it can be concluded that HBIM is the natural evolution of Digital Architectural Heritage or it can potentially be one of the main sources of DAH. Through the implementation of HBIM into the Digital Heritage field, it can be anticipated that HBIM will provide new opportunities for the DAH field, for example, HBIM applications can provide a new Virtual Reality (VR) system, which can allow the public to interact with the 3-D environments of these heritage sites (Aurel Schnabel and Aydin, 2015). Further, the ability of HBIM to store semantic inter-related information can be successfully applied to the cultural heritage sites that can be more easily managed (Garagnani and Manferdini, 2013). The resulting wider access to, and sharing of, information and databases allows the experts of cultural heritage (CH) to archive open, reliable, reusable, digital backups by themselves (Mudgea et al., 2007). The most important benefit can be found through the integration of the different data sources into a common model.

5.6   Limitations and challenges

There were many challenges facing the JHBIM project, in particular relating to the equipment, the historical area, and the case study, which will be discussed next.

Constrained resources and equipment availability challenges

- There is just one laser scanning device in Jeddah city, and it was busy with other projects at King Abdul Aziz University.
- During the on-site work, there was an issue with the laser scanning software, and it took more than a month to be repaired.
- The Cyclone's license was not active for all laptops in the Geomatics Department at KAU.
- The licenses for the BIM 360 Field and BIM 360 Glue were 30-day trial versions, which resulted in a loss of data.

Historical area

- Climatic condition challenges: the weather was hot (39°–43°) and the max humidity was 89% when the laser scanning survey took place on 10 August 2014.
- There was no care taken by some of the residents of Historic Jeddah and some of the municipality employees.
- A number of targets have been removed, which caused some issues with the registration.

The case study

- It took a long time to gain permission to scan the house.
- The location was not in good condition (lots of damage and garbage).
- There was a huge amount of architectural details.

Time-related challenges

- With regards to the time necessary to produce a professional 3-D model for each building with a sufficient level of detailing, each building needed almost one month for modelling.

5.7   Book contributions

The novel contributions made by this book for using HBIM as a new model for meeting the requirement of UNESCO's World Heritage nomination file (WHNF) are classified into the five main themes.

5.7.1   *A comprehensive review of UNESCO's WHNF: process, issues, and current solutions in the case of heritage buildings*

As with many systems in the world that have achieved significant recognition, and UNESCO's World Heritage nomination system being one of them, there are several challenges and issues being faced, and in this regard, policies and strategies are continually being adjusted in order to combat these challenges. Almost all of these processes have become much more difficult, requiring special effort in terms of building the capacity of people in the states' parties who have been charged with the implementation of the convention. This relates to the first part of the main question of the book: How can Heritage BIM provide innovation in creating missing information for UNESCO's World Heritage status? Therefore, answering this question was achieved via providing a review of UNESCO's WHNF in the case of heritage buildings. This addressed the main issues of the book, which includes the pieces of missing information during the nomination process, and the issues with the WHNF, as well as the proposed solutions for solving this matter.

5.7.2   *A review of BIM and the development toward Heritage BIM*

Research around BIM and the development toward Heritage BIM was required to investigate whether it would be useful for it to be employed in preparing UNESCO's World Heritage nomination file. While BIM began with the construction project's design process, Heritage BIM was developed in order to capture the existing construction to simulate them in a BIM form.

Through developing Jeddah Heritage BIM (the case study), Murphy's method of HBIM was examined and developed to be suitable for the case study and the BIM software (Autodesk Revit). Besides, Murphy (2012, p. 13) defined HBIM as "a novel prototype library of parametric objects, based on historic architectural data", as well as "a mapping system for modelling the objects library", based on the survey data of terrestrial laser scanning and image survey data. However, through employing Heritage BIM for the Historic Jeddah buildings, it has been noted that Heritage BIM was more than a parametric objects library. According to the British Standard BSI-PAS-1192 definition of BIM and the idea of BIM level 3 for asset management, it is a "process of designing, constructing or operating a building or infrastructure asset using electronic object-oriented information". Additionally, through focusing on the British Standard BSI (1192:2007), (1192–2:2013), and (1192–3:2014), it can be noted that the modern idea of BIM focuses more on the production of collaborative construction data and has been developed to include the operational stages throughout the lifecycle of the project (The British Standards Institution, 2014, 2013). Then, the project model is developed to include the idea of BIM level 3 through a Common Data Environment (CDE), which can provide access to the project data in different versions and at different levels of user access. Afterwards, this idea has also been developed to include non-graphical deliverables with the operational data, as well as the levels of model definition.

In the case of JHBIM, the parametric objects library supported the idea of BIM and also allowed for the speeding up the process of creating the Jeddah Heritage BIM model. Thus, via this book, a new definition of Heritage BIM has been developed, which defines Heritage BIM as, "a digital representation of the current situation of the physical and functional characteristics of the heritage buildings with respect to any modifications, restoration, and maintenance during the heritage building lifecycle." This can be based on laser scanning, photogrammetry, or any future advanced surveying technology.

5.7.3 Providing UNESCO's WHNF in a common and standard framework within Heritage BIM

This will be the first response to the main question, which is, *How can Heritage BIM provide innovation in creating missing information for UNESCO's World Heritage status? And what additional cultural value can a sustainable update of HBIM provide for such sites?* Furthermore, through focusing on section 3.3, the huge need for a new method is obvious, or the integration of almost all of the solutions in section 3.4 into one method. Thus, it is a necessity to come up with a common novel process in order to meet UNESCO's WHNF requirements and unite all efforts under one umbrella method. Heritage BIM can offer this common method to be shared, managed, and examined remotely, while increasing the accuracy of UNESCO's WHC decisions. In terms of section 2.2 and through analysing the UNESCO requirements, in the case of heritage buildings, it can be noted that the primary keys to solve these issues and to meet the requirements are through inputting information into an interactive database that can be remotely accessed and allows for the addition of comments and reports from different levels of users. With reference to section 2.4, employing Building Information Modelling (BIM) can meet these requirements in an interactive 3-D application with the ability to integrate different types of data such as CAD, GIS, GPS, laser scanning, images and much more data, which can be described as Big Data and can be well structured, such as with the metadata. Through inputting the information into the Heritage BIM process step by step as is required (the UNESCO WHNF requirements are mentioned in section 2.2), a common structure form containing the complete information that is necessary can be generated. Providing a common standard for the nomination files can improve the efficiency, quality, performance, and reduce errors during the nomination process, as well as increasing the possibility of reducing costs.

Employing Heritage BIM to satisfy the requirements of UNESCO's World Heritage nomination file is a new field to be investigated. Furthermore, the development of Jeddah Heritage BIM (JHBIM) to examine the concept of meeting the requirements of UNESCO's World Heritage nomination file through employing Heritage BIM was provided in chapter 4. This was the main contribution of this book.

Through employing Heritage BIM in the nomination process, it is possible to provide a framework for the organisations that are lacking in knowledge and resources to conduct assessments for potential World Heritage properties, as was demonstrated by Santana-Quintero and Van Balen (2009). As was mentioned in section 3.2.2, the issue of an absence of experience in relation to those involved in the nomination process and the unclear missions can be solved by employing Heritage BIM.

5.7.4   Developing a new method to produce complete engineering information regarding the heritage buildings in Jeddah through JHBIM

Employing BIM for the heritage buildings in Historic Jeddah is a new method for producing complete engineering information. Besides, JHBIM can support the conservation and maintenance management of the heritage buildings in Historic Jeddah. This can be achieved via the documentation of deteriorating aspects through using photogrammetry and 3-D laser scanning, obtaining high accuracy in documentation output, ensuring construction stability through the use of engineering modelling and total stations, as well as saving time and cost by using the technologies of heritage BIM data collectors. Furthermore, through JHBIM, the traditional building methods of the Hijazi buildings will be better understood. The JHBIM can act as an interactive database in the future through developing this database, which will include all the required data for any conservation or preservation projects in Historic Jeddah. One of the issues facing the heritage buildings currently is that there is no complete engineering information regarding these heritage buildings, which can be resolved by employing JHBIM.

5.7.5   Creating a heritage architecture objects library, which can be used in any heritage project in the Middle East

The Hijazi Architectural Objects Library (HAOL) was developed to assist with the process of creating the JHBIM model in several ways. While the modelling step of the heritage building can take more than six months to complete, the objects library can reduce that process to less than a month. HAOL reproduces Hijazi elements as 3-D computer models, which are modelled using a Revit Family (RFA). Since the HAOL is also dependent on the image survey and point cloud data, it is of high accuracy.

Additionally, the HAOL can recreate some of the architectural elements that are partially destroyed, in order to conserve or restore these unique aspects of these historical buildings. These Hijazi objects, such as Roshan and Mashrabiyah, have become the vocabulary of the Old Jeddah buildings. On the other hand, there is a huge gap in the Hijazi architectural library in providing these unique elements. Due to this issue, the Hijazi Architectural Objects Library has been created for the JHBIM project. The objects library of JHBIM can be connected to the data collected in a database, where each single modification of a parameter leads to a change in the shape of the object. As a result, considering the level of detail is very important; it is equally important that the object models can be simplified and easily modified in order to be suitable for the preservation plan, and for the models to have a greater opportunity to be utilised. This is particularly required for the buildings in Historic Jeddah, as the building objects are always unrivalled and irregular. The concept of this library is to be used as a plug-in for existing BIM application platforms, such as Autodesk Revit. The HAOL can be reused and inserted into different heritage buildings in Historic Jeddah or any similar heritage cities in the Middle East.

Bibliography

Abdelhafiz, A., 2009. *Integrating digital photogrammetry and terrestrial laser scanning*. Techn. Univ., Inst. für Geodäsie und Photogrammetrie.

Abmayr, T., Härtl, F., Reinköster, M., Fröhlich, C., 2005. Terrestrial laser scanning: Applications in cultural heritage conservation and civil engineering, in: Proceedings of the ISPRS Working Group V/4 Workshop 3D-ARCH 2005, Virtual Reconstruction and Visualization of Complex Architectures, International Archives of Photogrammetry, Remote Sensing and Spatial Information Sciences, Mestre-Venice.

Adan, A., Huber, D., 2011. 3D reconstruction of interior wall surfaces under occlusion and clutter, in: 3D Imaging, Modeling, Processing, Visualization and Transmission (3DIMPVT), 2011 International Conference on. IEEE, pp. 275–281.

Adas, A.A., 2013. Wooden bay window (Rowshan) conservation in Saudi-Hejazi heritage buildings. *ISPRS-Int. Arch. Photogramm. Remote Sens. Spat. Inf. Sci.* 1, 7–11.

Agarwal, S., Furukawa, Y., Snavely, N., Simon, I., Curless, B., Seitz, S.M., Szeliski, R., 2011. Building Rome in a day. *Commun. ACM* 54, 105–112.

Ahmad, A.M., 2014. *The use of refurbishment, flexibility, standardisation and BIM to support the design of a change-ready healthcare facility.* © Ahmad Mohammad Ahmad.

Alawi, I., 2013. Saudi Gazette: Old Jeddah Hit by Worst Fire in Years [WWW Document]. URL www.saudigazette.com.sa/index.cfm?method=home.regcon&contentID= 2010030465281 (accessed 10.21.13).

Al-Fakahani, H., 2005. *Jeddah: The bridge of the red sea: Progress and development.* The Arab Publishing House for Encyclopedias, Jeddah, KSA.

Alitany, A., Redondo Domínguez, E., & Adas, A., 2013a. The 3D documentation of projected wooden windows (The Roshans) in the old city of Jeddah (Saudi Arabia) using image-base techniques. *ISPRS Ann. Photogramm. Remote Sens. Spatial Inf. Sci., II-5/W1. XXIV International CIPA Symposium*, 7–12.

Alitany, A., Redondo, E., Fonseca, D., Riera, A.S., 2013b. Hybrid-ICT: Integrated methodologies for heritage documentation: Evaluation of the combined use of digital photogrammetry, 3D modeling and augmented reality in the documentation of architectural heritage elements, in: Information Systems and Technologies (CISTI), 2013 8th Iberian Conference on. IEEE, pp. 1–7.

Al-Lyaly, S.M.Z., 1990. *The traditional house of Jeddah: A study of the interaction between climate, form and living patterns.* University of Edinburgh.

Amm Mahfooz., 2011. *Local contractor at Old Jeddah* [Personal communication].

Amm Saad., 2011. *Local contractor at Old Jeddah* [Personal communication].

Andrews, D., Bedford, J., Blake, B., Bryan, P., Papworth, H., Cromwell, T., 2013. *Measured and drawn: Techniques and practice for the metric survey of historic buildings.* English Heritage.

Angawi, S., 1988. *Bayt Al-shafi restoration*. AMAR Headquarters, Jeddah, Saudi Arabia.

Aouad, G., Lee, A., Wu, S., 2006. *Constructing the future: nD modelling*. Routledge, Abingdon, UK.

Arab News, 2011. UNESCO refuses to consider Old Jeddah a world heritage site.

Arayici, Y., 2008. Towards building information modelling for existing structures. *Struct. Surv.* 26, 210–222.

Arayici, Y., & Aouad, G., 2010. Building information modelling (BIM) for construction lifecycle management. *Construction and Building: Design, Materials, and Techniques, 2010*, 99–118.

Arayici, Y., Khosrowshahi, F., Ponting, A.M., & Mihindu, S.A., 2009. Towards implementation of building information modelling in the construction industry. *Proceedings of the Fifth International Conference on Construction in the 21st Century: Collaboration and Integration in Engineering, Management and Technology*, 1342–1351.

Attar, R., Prabhu, V., Glueck, M., Khan, A., 2010. 210 king street: A dataset for integrated performance assessment, in: Proceedings of the 2010 Spring Simulation Multiconference. Society for Computer Simulation International, p. 177.

Aurel Schnabel, M., Aydin, S., 2015. BIM & Digital Heritage [WWW Document]. URL http://caadria2016.crida.net/exhibitors/bim-digital-heritage/ (accessed 4.9.16).

Autodesk, 2013a. BIM 360 Glue | Cloud-Based BIM Collaboration Software | Autodesk [WWW Document]. URL http://usa.autodesk.com/adsk/servlet/pc/index?id=21318325&siteID=123112 (accessed 6.25.13).

Autodesk, 2013b. Autodesk BIM 360 Field® [WWW Document]. URL www.autodesk-bim360.com/bim-360-field-overview (accessed 12.16.13).

Autodesk, 2016a. About Autodesk Revit DB Link [WWW Document]. URL http://help.autodesk.com/view/RVT/2016/ENU/?guid=GUID-639C5DB8-3AB7-4803-9EFE-72CA4F714F69 (accessed 7.4.16).

Autodesk, 2016b. A360: Home [WWW Document]. URL https://a360.autodesk.com/ (accessed 9.29.16).

Autodesk Inc, 2015. Software for Building Design and Construction [WWW Document]. Autodesk. URL www.autodesk.com/products/revit-family/features/all/gallery-view

Aziz, K.A., Siang, T.G., 2014. Virtual reality and augmented reality combination as a holistic application for heritage preservation in the UNESCO world heritage site of Melaka. *Int. J. -Cial Sci. Humanity* 4, 333–338.

Backes, D., Thomson, C., Malki-Epshtein, L., & Boehm, J., 2014. *Chadwick GreenBIM: Advancing operational understanding of historical buildings with BIM to support sustainable use*. Building Simulation and Optimization Conference, London, UK.

Baik, A., 2016. Documentation of the Nasif Historical House, in historical Jeddah, Saudi Arabia, using terrestrial laser scanning and image survey methods, in: Proceedings of the Eighth Saudi Students Conference in the UK. World Scientific, pp. 767–780.

Baik, A., Alitany, A., Boehm, J., Robson, S., 2014. Jeddah historical building information modelling "JHBIM": Object library. *ISPRS Ann. Photogramm. Remote Sens. Spat. Inf. Sci.* 2, 41.

Baik, A., Boehm, J., & Robson, S., 2013. Jeddah Historical Building Information Modeling 'JHBIM' Old Jeddah—Saudi Arabia. *ISPRS – International Archives of the Photogrammetry, Remote Sensing and Spatial Information Sciences*, XL-5/W2, 73–78. https://doi.org/10.5194/isprsarchives-XL-5-W2-73-2013.

Baik, A., Yaagoubi, R., Boehm, J., 2015. Integration of Jeddah historical BIM and 3D GIS for documentation and restoration of historical monument, in: ISPRS-International Archives of the Photogrammetry, Remote Sensing and Spatial Information Sciences, pp. 29–34.

Bandarin, F., 2007. Introduction: Present and future challenges to the world heritage convention. World Heritage—Challenges Millenn, pp. 18–24.

Barazzetti, L., Banfi, F., Brumana, R., Oreni, D., Previtali, M., Roncoroni, F., 2015. HBIM and augmented information: Towards a wider user community of image and range-based reconstructions. *ISPRS-Int. Arch. Photogramm. Remote Sens. Spat. Inf. Sci.* 1, 35–42.

Beraldin, J.-A., 2004. Integration of laser scanning and close-range photogrammetry-the last decade and beyond, in: International Society for Photogrammetry and Remote Sensing.

BIM Task Group, 2013. COBie UK 2012 [WWW Document]. BIMTaskGroup.org. URL www.bimtaskgroup.org/cobie-uk-2012/

Boehler, W., Vicent, M.B., Marbs, A., 2003. Investigating laser scanner accuracy. Presented at the XIXth Cipa Symposium.

Boeykens, S., 2011. Using 3D design software, BIM and game engines for architectural historical reconstruction. *Des. Together-CAADfutures* 493–509.

Boeykens, S., Himpe, C., Martens, B., 2012. A case study of using BIM in historical reconstruction: The Vinohrady Synagogue in Prague. *Digit. Phys. Phys. Digit.* 729–738.

Böhm, J., Haala, N., Becker, S., 2007. Façade modelling for historical architecture, in: XXI International CIPA Symposium, pp. 1–6.

Bokhari, A. 2006. Conservation in the Historic District of Jeddah. King Saud University, Riyadh, Saudi Arabia.

Bolla, G., Batisse, M., Bolla, G., 2005. Episodes of a painstaking gestation. Invent. "world Heritage" Hist. Pap. UNESCO Action Seen Protag. Witn.

Boon, J., Prigg, C., 2012. Evolution of quantity surveying practice in the use of BIM: The New Zealand experience, in: Joint CIB International Symposium of W055, W065, W089, W118.

The British Standards Institution, 2013. *PAS 1192–2:2013 specification for information management for the capital/delivery phase of construction projects using building information modelling*, London, BSI Standards Limited.

The British Standards Institution, 2014. *PAS 1192–3:2014 specification for information management for the operational phase of assets using building information modelling*, London, BSI Standards Limited.

Brumana, R., 1990. Sant'Ambrogio's Basilica in Milan: A study on photogrammetric surveys in the S. Vittore in Ciel d'Oro's dome, in: Society of Photo-Optical Instrumentation Engineers (SPIE) Conference Series, p. 908.

Brumana, R., Crippa, B., Vassena, G., 1990. Analytical treatment and description of the altimetric check of the St. Marcus' Basilica in Venice, in: Society of Photo-Optical Instrumentation Engineers (SPIE) Conference Series, p. 166.

BSI, 2007. *Collaborative production of architectural, engineering and construction information: Code of practice*, London, BSI Standards Limited.

Budroni, A., Boehm, J., 2010. Automated 3D reconstruction of interiors from point clouds. *Int. J. Archit. Comput.* 8, 55–73.

Budroni, A., Böhm, J., 2010. Automatic 3D modelling of indoor Manhattan-world scenes from laser data, in: ISPRS Symp. Close Range Image Measurement Techniques.

Büyüksalih, G., Jacobsen, K., 2006. Comparison of DEM generation by very high resolution optical satellites. EARSeL Wars. Pol.

Campbell, R.J., Flynn, P.J., 2001. A survey of free-form object representation and recognition techniques. *Comput. Vis. Image Underst.* 81, 166–210.

Cantzler, H., 2003. *Improving architectural 3D reconstruction by constrained modelling*. University of Edinburgh. College of Science and Engineering.

Cheng, H.-M., Yang, W.-B., Yen, Y.-N., 2015. BIM applied in historical building documentation and refurbishing. *Int. Arch. Photogramm. Remote Sens. Spat. Inf. Sci.* 40, 85.

Cheng, X.J., Jin, W., 2006. Study on reverse engineering of historical architecture based on 3D laser scanner. In *Journal of Physics: Conference Series*. IOP Publishing, p. 843.

Cheok, G.S., Stone, W.C., Lipman, R.R., Witzgall, C., 2000. Ladars for construction assessment and update. *Autom. Constr.* 9, 463–477.

Chevrier, C., Charbonneau, N., Grussenmeyer, P., Perrin, J.-P., 2010. Parametric documenting of built heritage: 3D virtual reconstruction of architectural details. *Int. J. Archit. Comput.* 8, 135–150.

Cho, H., Lee, K.H., Lee, S.H., Lee, T., Cho, H.J., Kim, S.H., Nam, S.H., 2011. Introduction of construction management integrated system using BIM in the Honam highspeed railway lot no. 4–2, in: Proc. 28th ISARC Seoul Korea.

Climatemps.com, 2015. Jeddah Climate Jeddah Temperatures Jeddah Weather Averages [WWW Document]. URL www.jeddah.climatemps.com/ (accessed 1.1.16).

Dai, F., Lu, M., Kamat, V.R., 2010. Analytical approach to augmenting site photos with 3D graphics of underground infrastructure in construction engineering applications. *J. Comput. Civ. Eng.* 25, 66–74.

D'Andrea, A., Niccolucci, F., Bassett, S., Fernie, K., 2012. 3D-ICONS: World heritage sites for Europeana: Making complex 3D models available to everyone, in: Virtual Systems and Multimedia (VSMM), 2012 18th International Conference on. IEEE, pp. 517–520.

Darie, D., 2014. *Building information modelling for museums* (MSC in Surveying). University College London, London, UK.

Del Giudice, M., Osello, A., 2013. BIM for cultural heritage. *Int. Arch. Photogramm. Remote Sens. Spat. Inf. Sci.* 40, 225–229.

De Luca, L., Driscu, T., Peyrols, E., Labrosse, D., Berthelot, M., 2014. A complete methodology for the virtual assembling of dismounted historic buildings. *Int. J. Interact. Des. Manuf. IJIDeM* 8, 265–276.

de Merode, E., Smeets, H.J., Westrik, C., 2004. Linking universal and local values: Managing a sustainable future for world heritage: A conference organized by the Netherlands national commission for UNESCO, in: Collaboration with the Netherlands Ministry of Education, Culture and Science, 22–24 May 2003. UNESCO World Heritage Centre.

Dickinson, J., Pardasani, A., Ahamed, S.S., Kruithof, S., 2009. A survey of automation technology for realizing as-built models of services. In: Belloni, K., Kojima, J., Seppä, I.P. (Eds.), *Presented at the CIB IDS 2009: Improving construction and use through integrated design solutions*. VTT Technical Research Centre of Finland, Helsinki, Finland, pp. 365–381.

Dijkgraaf, C., 2003. How world heritage sites disappear: Four cases, four threats. Link. Univers. Local Values 32.

Dirix, E., 2015. Former places of worship in Dublin: Historic building information. University of Leuven, Belgium.

Donath, D., Petzold, F., Braunes, J., Fehlhaber, D., Tauscher, H., Junge, R., Göttig, R., 2010. *IT-gestützte projekt-und zeitbezogene Erfassung und Entscheidungsunterstützung in der frühen Phase der Planung im Bestand (Initiierungsphase) auf Grundlage eines IFC-basierten CMS*. Fraunhofer-IRB-Verlag.

Dore, C., Murphy, M., 2012. Integration of Historic Building Information Modeling (HBIM) and 3D GIS for recording and managing cultural heritage sites, in: Virtual Systems and Multimedia (VSMM), 2012 18th International Conference on. IEEE, pp. 369–376.

Dore, C., Murphy, M., 2013. Semi-automatic modelling of building facades with shape grammars using historic building information modelling. *ISPRS Int. Arch. Photogramm. Remote Sens. Spat. Inf. Sci.* 40, 5.

Dumper, M., Larkin, C., 2012. The politics of heritage and the limitations of international agency in contested cities: A study of the role of UNESCO in Jerusalem's old city. *Rev. Int. Stud.* 38, 25–52.

Dunston, P.S., Wang, X., 2005. Mixed reality-based visualization interfaces for architecture, engineering, and construction industry. *J. Constr. Eng. Manag.* 131, 1301–1309.

Eastman, C.M., 1975. The use of computers instead of drawings in building design. *AIA J.* 63, 46–50.

Eastman, C., 2006. *Report on integrated practice university and industry research in support of BIM.* American Institute of Architects, Georgia Institute of Technology.

Eastman, C.M., Teicholz, P., Sacks, R., Liston, K., 2011. *BIM handbook: A guide to Building Information Modeling for owners, managers, designers, engineers and contractors.* Wiley.com.

eBIM, 2015. Heritage BIM: eBIM [WWW Document]. URL http://ebim.co.uk/heritage-bim/ (accessed 4.10.16).

ECTURAE, I. design group., 2012. Al Balad Historic District Survey [Eng]. TECTURAE Italian design group, Jeddah, Saudi Arabia.

Eleish, A., 2009. Heritage conservation in Saudi Arabia, in: Proc. Jt. Int. Symp. IAPS-CSBE Hous. Netw. Revital. Built Environ. Re-Qualif. Old Places New Uses.

El Garouani, A., Alobeid, A., 2013. Digital surface model generation for 3D city modeling (Fez, Morocco), in: Eighth National GIS Symposium in Saudi Arabia, "The Road For Building Saudi Arabia GIS" Organized by the High Committee of the GIS at the Eastern Province.

El-Hakim, S., Beraldin, J.-A., Gonzo, L., Whiting, E., Jemtrud, M., & Valzano, V. (2005). A hierarchical 3D reconstruction approach for documenting complex heritage sites. *Proceedings of the XX CIPA International Symposium, Torino, Italy,* 26.

Fai, S., Filippi, M., Paliaga, S., 2013. Parametric modelling (BIM) for the documentation of vernacular construction methods: A BIM model for the commissariat building, Ottawa, Canada, in: International CIPA Symposium, Vol. II-5/W1, Strasbourg, France, 2–6 September.

Fai, S., Graham, K., Duckworth, T., Wood, N., Attar, R., 2011a. Building information modelling and heritage documentation, in: Paper presented to XXIII CIPA International Symposium, Prague, Czech Republic, 12–16 September.

Fai, S., Graham, K., Duckworth, T., Wood, N., Attar, R., 2011b. Building information modelling and heritage documentation, in: Proceedings of the 23rd International Symposium, International Scientific Committee for Documentation of Cultural Heritage (CIPA), Prague, Czech Republic, pp. 12–16.

Fai, S., Rafeiro, J., 2014. Establishing an appropriate Level of Detail (LoD) for a Building Information Model (BIM): West block, Parliament Hill, Ottawa, Canada. *ISPRS Ann. Photogramm. Remote Sens. Spat. Inf. Sci.* II-5, 123–130. doi:10.5194/isprsannals-II-5-123-2014

Fassi, F., Achille, C., Mandelli, A., Rechichi, F., Parri, S., 2015. A new idea of bim system for visualization, web sharing and using huge complex 3D models for facility management. *Int. Arch. Photogramm. Remote Sens. Spat. Inf. Sci.* 40, 359.

Feilden, B., 2012. *Conservation of historic buildings.* Routledge, Abingdon, UK.

Fitzgibbon, A.W., Eggert, D.W., Fisher, R.B., 1997. High-level CAD model acquisition from range images. *Comput.-Aided Des.* 29, 321–330.

Foster and Partners, 2015. New Transport System for Jeddah | Foster + Partners [WWW Document]. URL www.fosterandpartners.com/news/archive/2015/03/foster-partners-appointed-to-design-new-transport-system-for-jeddah/ (accessed 9.25.16).

Frahm, J.-M., Pollefeys, M., Lazebnik, S., Gallup, D., Clipp, B., Raguram, R., Wu, C., Zach, C., Johnson, T., 2010. Fast robust large-scale mapping from video and internet photo collections. *ISPRS J. Photogramm. Remote Sens.* 65, 538–549.

Fregonese, L., Achille, C., Adami, A., Fassi, F., Spezzoni, A., Taffurelli, L., 2015. BIM: An integrated model for planned and preventive maintenance of architectural heritage, in: 2015 Digital Heritage. IEEE, pp. 77–80.

Furukawa, Y., Ponce, J., 2010. Accurate, dense, and robust multiview stereopsis, in: Pattern Anal. Mach. Intell. IEEE Trans. On 32, pp. 1362–1376.

Fussel, T., Beazley, S., Aranda-Mena, G., Chevez, A., Crawford, J., Succar, B., Drogemuller, R., Gard, S., Nielsen, D., 2009. *National guidelines for digital modelling: Case studies*. CRC for Construction Innovation.

Garagnani, S., Manferdini, A.M., 2013. Parametric accuracy: Building Information Modeling process applied to the cultural heritage preservation, in: ISPRS-International Archives of the Photogrammetry, Remote Sensing and Spatial Information Sciences, pp. 87–92.

Gillot, L., Del, A., 2011. Preparation and submission of the nomination file of the Oasis of Figuig (Morocco) for inscription on the world heritage list: Impacts and uses of a GIS. *Geoinformatics FCE CTU* 6, 140–148.

Graphisoft SE, 2015. Archicad [WWW Document]. Graphisoft. URL www.graphisoft.com/archicad/

Greenlaw, J., 1995. *The coral buildings of Suakin, Islamic architecture, planning, design and domestic arrangements in a Red Sea port*, 2ed ed. Kegan Paul International Limited, London, New York.

Group, B.I.W., 2011. A Report for the Government Construction Client Group Building Information Modelling (BIM) Working Party Strategy Paper. Cabinet Office. URL www.bimtaskgroup.org/wp-content/uploads/2012/03/BIS-BIM-strategy-Report.pdf

Gupta, A., 2013. *Developing a BIM-based methodology to support renewable energy assessment of buildings*. Cardiff University, Cardiff, UK.

Guttentag, D.A., 2010. Virtual reality: Applications and implications for tourism. *Tour. Manag.* 31, 637–651.

Haala, N., Anders, K.-H., 1996. Fusion of 2D-GIS and image data for 3D building reconstruction. *Int. Arch. Photogramm. Remote Sens.* 31, 285–290.

Hajian, H., Becerik-Gerber, B., 2010. Scan to BIM: Factors affecting operational and computational errors and productivity loss, in: 2010 Proceedings of the 27th International Symposium on Automation and Robotics in Construction. IAARC, Bratislava, Slovakia, pp. 265–272.

Hanke, K., Grussenmeyer, P., 2002. Architectural photogrammetry: Basic theory, procedures, tools, in: ISPRS Commission.

Heritage, E., 2011. *3D laser scanning for heritage: Advice and guidance to users on laser scanning in archaeology and architecture*, 2nd ed. English Heritage, York, UK.

Hichri, N., Stefani, C., De Luca, L., Veron, P., 2013a. Review of the "as-built BIM" approaches. *ISPRS Arch. Photogramm Remote Sens Spat. Inf.* 107–112.

Hichri, N., Stefani, C., Luca, L.D., Veron, P., 2013b. Review of the "as-built BIM" approaches. In Boehm, J., Remondino, F., Kersten, T., Fuse, T., Gonzalez-Aguilera, D. (Eds.), *International society for photogrammetry and remote sensing*. International Society for Photogrammetry and Remote Sensing, Trento, Italy, pp. 107–112.

Hirschmuller, H., 2005. Accurate and efficient stereo processing by semi-global matching and mutual information, in: Computer Vision and Pattern Recognition, 2005. CVPR 2005. IEEE Computer Society Conference on. IEEE, pp. 807–814.

Huber, D., Akinci, B., Tang, P., Adan, A., Okorn, B., Xiong, X., 2010. Using laser scanners for modeling and analysis in architecture, engineering, and construction, in: Information Sciences and Systems (CISS), 2010 44th Annual Conference on. IEEE, pp. 1–6.

ICOMOS, 2013. ICOMOS Shared Built Heritage: Cultural Heritage Connections [WWW Document]. URL www.culturalheritageconnections.org/wiki/ICOMOS_Shared_Built_Heritage (accessed 3.7.16).

Ikeuchi, K., 2001. Modeling from reality, in: 3-D Digital Imaging and Modeling, 2001. Proceedings. Third International Conference on. IEEE, pp. 117–124.

Ilter, D., Ergen, E., 2015. BIM for building refurbishment and maintenance: Current status and research directions. *Struct. Surv.* 33, 228–256.

Jayakody, A., Rupasinghe, L., Perera, K., Herath, H., Thennakoon, T. M. A., & Premanath, S. U. 2013. The Development of the CityGML GeoBIM Extension for Real-Time Assessable Model (Integration of BIM and GIS). The Development of the CityGML–PNCTM, Vol. 2, 109–113.

Jokilehto, J., 2015. What is OUV? Defining the Outstanding Universal Value of Cultural World Heritage Properties. Monum. Sites 16.

Kiviniemi, A., 2008. *Review of the development and implementation of IFC compatible BIM.* Erabuild.

Kiviniemi, A., 2011. The effects of Integrated BIM in processes and business models. *Distrib. Intell. Des.* 125–135.

Klein, L., Li, N., Becerik-Gerber, B., 2012. Imaged-based verification of as-built documentation of operational buildings. *Autom. Constr.* 21, 161–171.

Kolbe, T.H., Gröger, G., Plümer, L., 2008. CityGML: 3D city models and their potential for emergency response. *Geospatial Inf. Technol. Emerg. Response* 257.

Kwoh, L.K., Liew, S.C., Xiong, Z., 2004. Automatic DEM generation from satellite image, in: The 25 Th Asian Conference & 1 Th Asain Space Conference on Remote Sensing, November, pp. 22–26.

Lee, S.H., Park, S.I., Park, J., Seo, K.W., 2014. Open BIM-based information modeling of railway bridges and its application concept, in: 2014 International Conference on Computing in Civil and Building Engineering, pp. 23–25.

Leica Geosystems, 2013. Leica ScanStation C10 [WWW Document]. URL http://hds.leica-geosystems.com/en/Leica-ScanStation-C10_79411.htm (accessed 10.24.13).

Leica Geosystems, 2006. *Inc. HDS training manual: Scanning & cyclone 5.4. 1.* Leica Geosystems Inc, San Ramon, CA.

Leite, F., Akcamete, A., Akinci, B., Atasoy, G., Kiziltas, S., 2011. Analysis of modeling effort and impact of different levels of detail in building information models. *Autom. Constr.* 20, 601–609.

Letellier, R., Eppich, R., 2015. *Recording, documentation and information management for the conservation of heritage places.* Routledge, Abingdon, UK.

Linaza, M.T., Juaristi, M., Garcia, A., 2014. Reusing multimedia content for the creation of interactive experiences in cultural institutions. In *3D research challenges in cultural heritage.* Springer, pp. 104–118.

Lipp, M., Wonka, P., Wimmer, M., 2008. Interactive visual editing of grammars for procedural architecture. ACM *Trans. Graph. TOG* 27, 102.

Liu, X., Akinci, B., 2009. Requirements and evaluation of standards for integration of sensor data with building information models. *Comput. Civ. Eng.* 10.

Logothetis, S., Delinasiou, A., Stylianidis, E., 2015. Building Information Modelling for cultural heritage: A review. *ISPRS Ann. Photogramm. Remote Sens. Spat. Inf. Sci.* 1, 177–183.

Lorenzini, M. 2009. Semantic approach to 3D historical reconstruction. Proceedings of the 3rd ISPRS International Workshop 3D-ARCH 2009:" 3D Virtual Reconstruction and Visualization of Complex Architectures" Trento, Italy, 25–28 February 2009, 38–43.

Luke, C., Kersel, M.M., 2012. *Soft power, hard heritage: US cultural diplomacy and archaeology.* Routledge, London.

Ma, Y.P., Hsu, C.C., Lin, M.C., Tsai, Z.W., Chen, J.Y., 2015. Parametric workflow (BIM) for the repair construction of traditional historic architecture in Taiwan. *Int. Arch. Photogramm. Remote Sens. Spat. Inf. Sci.* 40, 315.

Macher, H., Landes, T., Grussenmeyer, P., Alby, E., 2014. Semi-automatic segmentation and modelling from point clouds towards historical building information modelling, in: *Digital heritage: Progress in cultural heritage: Documentation, preservation, and protection.* Springer, pp. 111–120.

Marshall, D., 2011. *Preparing world heritage nominations.* UNESCO.

Matthew, R., Johnson-Marshall, 1980. *Jeddah: Historic area study, design demonstration study.* Municipality of Jeddah, Jeddah, Saudi Arabia.

Mechelke, K., Kersten, T.P., Lindstaedt, M., 2007. Comparative investigations into the accuracy behaviour of the new generation of terrestrial laser scanning systems. *Proc Opt.* 319–327.

Meskell, L., Liuzza, C., Bertacchini, E., Saccone, D., 2015. Multilateralism and UNESCO world heritage: Decision-making, states parties and political processes. *Int. J. Herit. Stud.* 21, 423–440.

Mills, J., Barber, D., 2004. Geomatics techniques for structural surveying. *J. Surv. Eng.* 130, 56–64.

Moore, M., 2001. Conservation documentation and the implications of digitisation. *J. Conserv. Mus. Stud.* 7.

Moriwaki, A., 2014. Building Lifecycle Management Fosters a BIM Level 3 Approach for End-to-End AEC Collaboration [Whitepaper] | White Papers | the BIM Hub [WWW Document]. URL https://thebimhub.com/2014/10/11/building-lifecycle-management-fosters-a-bim-leve-3/#.VwfTgUZUrrS (accessed 4.8.16).

Mudgea, M., Ashleyb, M., Schroer, C., 2007. A digital future for cultural heritage. Available HttpculturalheritageimagingOrgWhatWeDoPublicationscipa2007CIPA2007 Pdf.

Murphy, M., 2012. *Historic Building Information Modelling (HBIM) for Recording and Documenting Classical Architecture in Dublin 1700 to 1830* (Doctor of Philosophy thesis). Trinity College Dublin, Dublin.

Murphy, M., 2013. Historic Building Information Modelling: Adding intelligence to laser and image based surveys of European classical architecture. *ISPRS J. Photogramm. Remote Sens.* 76, 89–102.

Murphy, M., McGovern, E., Pavia, S., 2009. Historic Building Information Modelling (HBIM). *Struct. Surv.* 27, 311–327. doi:10.1108/02630800910985108

Murphy, M., McGovern, E., Pavia, S., 2013. Historic Building Information Modelling: Adding intelligence to laser and image based surveys of European classical architecture. *ISPRS J. Photogramm. Remote Sens.* 76, 89–102. doi:10.1016/j.isprsjprs.2012.11.006

Navrud, S., Ready, R.C., 2002. *Valuing cultural heritage: Applying environmental valuation techniques to historic buildings, monuments and artifacts.* Edward Elgar Publishing.

Nawwar, S., 2013. *Jeddah historic preservation department.* Jeddah Municipality.

NBIMS-US, 2015. About the National BIM Standard-United States [WWW Document]. www.nationalbimstandard.org/. URL www.nationalbimstandard.org/faq.php#faq1

NBS, 2015. Building Information Modelling [WWW Document]. BIM Levels Explain. URL www.thenbs.com/topics/bim/articles/bim-levels-explained.asp

Nex, F., Rinaudo, F., 2008. Multi-image matching: An "old and new" photogrammetric answer to lidar techniques. *Proc. Int. Arch. Photogramm. Remote Sens. Spat. Inf. Sci. IAPRSSIS* 37, 621–626.

NHBC, 2013. *Building Information Modelling: An introduction for house builders.* BSRIA Limited.

Nüchter, A., Hertzberg, J., 2008. Towards semantic maps for mobile robots. *Robot. Auton. Syst.* 56, 915–926.

Orbasli, A., 2007. Training conservation professionals in the Middle East. *Built Environ.* 33, 307–322.

Oreni, D., 2013. From 3D content models to HBIM for conservation and management of built heritage, in: Computational Science and Its Applications—ICCSA 2013, 0302-9743. Springer Berlin Heidelberg, pp. 344–357.

Oreni, D., Brumana, R., Banfi, F., Bertola, L., Barazzetti, L., Cuca, B., Previtali, M., Roncoroni, F., 2014. Beyond crude 3D models: From point clouds to historical Building Information Modeling via NURBS. In: *Digital heritage: Progress in cultural heritage: Documentation, preservation, and protection*. Springer, pp. 166–175.

Oreni, D., Brumana, R., Cuca, B., 2012. Towards a methodology for 3D content models: The reconstruction of ancient vaults for maintenance and structural behaviour in the logic of BIM management, in: Virtual Systems and Multimedia (VSMM), 2012 18th International Conference on. IEEE, pp. 475–482.

Oreni, D., Brumana, R., Cuca, B., Georgopoulos, A., 2013. HBIM for conservation and management of built heritage: Towards a library of vaults and wooden bean floors, in: CIPA 2013XXV International Symposium, ISPRS Annals, pp. 1–6.

Oreni, D., Brumana, R., Della Torre, S., Banfi, F., Barazzetti, L., Previtali, M., 2014. Survey turned into HBIM: The restoration and the work involved concerning the Basilica di Collemaggio after the earthquake (L'Aquila). ISPRS Ann. Photogramm. Remote Sens. Spat. Inf. Sci. Riva Garda Italy, June 23–25.

Owen-John, H., 2015, October 20. *Developing a new method to prepare the UNESCO World Heritage nomination file* [Interview].

Pannell, S., 2006. Reconciling nature and culture in a global context, in: Lessons World Herit. List Coop. Res. Cent. Trop. Rainfor. Ecol. Manag. Rainfor. CRC Cairns Aust.

Park, C.-S., Lee, D.-Y., Kwon, O.-S., Wang, X., 2013. A framework for proactive construction defect management using BIM, augmented reality and ontology-based data collection template. *Autom. Constr.* 33, 61–71.

Paul, A., 2013. *Renaissance revit: Creating classical architecture with modern software*, 1B ed. CreateSpace Independent Publishing Platform, USA.

Penttilä, H., Rajala, M., Freese, S., 2007. Building Information Modelling of modern historic buildings. *Predict. Future 25th ECAADe Konf. Frankf. Am Main Ger.* 607–613.

Pesce, A., 1974. *Jiddah: Portrait of an Arabian city*. ICON Group International.

Pfeifer, N., Böhm, J., 2008. *Advances in photogrammetry, remote sensing and spatial information sciences, 2008 ISPRS congress book*, Vol. 7, ISPRS. Beijing, pp. 169–184.

Pike, B., 2012. DIY FM: Bridging the AEC/FM gap using revit DB link, ASP.NET, HTML, VB and Java, in: Presented at the Technology Conference North America 2012.

Pollefeys, M., Nistér, D., Frahm, J.-M., Akbarzadeh, A., Mordohai, P., Clipp, B., Engels, C., Gallup, D., Kim, S.-J., Merrell, P., Salmi, C., Sinha, S., Talton, B., Wang, L., Yang, Q., Stewénius, H., Yang, R., Welch, G., Towles, H., 2008. Detailed real-time urban 3D reconstruction from video. *International Journal of Computer Vision*, 10.

Ragette, F., 2003. *Traditional domestic architecture of the Arab region*. Edition Axel Menges.

Rajala, M., Penttilä, H., 2006. Testing 3D building modelling framework in building renovation, in: Communicating Space (S), Proceedings of the 24th Education and Research in Computer Aided Architectural Design in Europe Conference. University of Thessaly: Volos, Greece, pp. 268–275.

Rajendra, M.Y., Mehrotra, S.C., Kale, K.V., Manza, R.R., Dhumal, R.K., Nagne, A.D., Vibhute, A.D., 2014. Evaluation of partially overlapping 3D point cloud's registration by using ICP variant and cloudcompare. *ISPRS-Int. Arch. Photogramm. Remote Sens. Spat. Inf. Sci.* 1, 891–897.

Rao, K., 2010. A new paradigm for the identification, nomination and inscription of properties on the World Heritage List. *Int. J. Herit. Stud.* 16, 161–172.

Remondino, F., 2003. From point cloud to surface: The modeling and visualization problem, in: International Workshop on Visualization and Animation of Reality-Based 3D Models, p. 5.

Remondino, F., El-Hakim, S., Girardi, S., Rizzi, A., Benedetti, S., Gonzo, L., 2009. 3D virtual reconstruction and visualization of complex architectures: The "3D-ARCH" pProject, in: Proceedings of the ISPRS Working Group V/4 Workshop 3D-ARCH "Virtual Reconstruction and Visualization of Complex Architectures", 2009.

Remondino, F., Gaiani, M., Apollonio, F., Ballabeni, A., Ballabeni, M., Morabito, D., 2016. 3D documentation of 40 kilometers of historical porticoes-the challenge. *ISPRS Arch. Photogramm. Remote Sens. Spat. Inf. Sci.* 5, 711–718.

Remondino, F., Niederoest, J., 2004. Generation of high-resolution mosaic for photorealistic texture-mapping of cultural heritage 3D models, in: Proceedings of the 5th International Conference on Virtual Reality, Archaeology and Intelligent Cultural Heritage. Eurographics Association, pp. 85–92.

ReviverSoft, 2013. RFA File Extension [WWW Document]. URL www.reviversoft.com/file-extensions/rfa (accessed 5.25.16).

Riccio, F., 2014. Italian UNESCO world heritage: Forms of protection and management experiences, in: Almatourism: Journal of Tourism, Culture and Territorial Development. Scientific and Didactical Campus of Rimini and the Advanced School of Tourism Sciences of Alma Mater Studiorum, pp. 26–31.

Runne, H., Niemeier, W., Kern, F., 2001. Application of laser scanners to determine the geometry of buildings. In Grün, A., Kahmen, H. (Eds.), *Optical 3D measurement techniques V: Department of applied and engineering geodesy, institute of deodesy and geophysics.* Vienna University of Technology, Vienna, Austria, pp. 41–48.

Santana-Quintero, M., Blake, B., Eppich, R., Ouimet, C., 2008. Heritage documentation for conservation: Partnership in learning. In *The spirit of place.* ICOMOS.

Santana-Quintero, M., Van Balen, K., 2009. Rapid and cost-effective assessment for world heritage nominations, in: Proceedings of the CIPA XXII International Symposium: Digital Documentation, Interpretation & Presentation of Cultural Heritage. Ritsumeikan University.

Santos, P., Serna, S.P., Stork, A., Fellner, D., 2014. The potential of 3D internet in the cultural heritage domain. In *3D research challenges in cultural heritage.* Springer, pp. 1–17.

Saudi SCTA, 2015. *The workshop of mechanisms to preserve the architectural heritage.* Jeddah, Saudi Arabia.

Saygi, G., Agugiaro, G., Hamamcıoğ lu-Turan, M., Remondino, F., 2013. Evaluation of GIS and BIM roles for the information management of historical buildings, in: ISPRS Ann. Photogramm. Remote Sens. Spat. Inf. Sci. Vol. II-5W1 2013 XXIV Int. CIPA Symp. 2–6 September, Strasbg. Fr, pp. 283–288.

Saygi, G., Remondino, F., 2013. Management of architectural heritage information in BIM and GIS: State-of-the-art and future perspectives. *Int. J. Herit. Digit. Era* 2, 695–714.

Schmitt, T.M., 2009. Global cultural governance: Decision-making concerning world heritage between politics and science. *Erdkunde* 103–121.

SCTA, 2013. *Historic jeddah, the gate to makkah.* Saudi Commission for Tourism and Antiquities, Saudi Arabia, Jeddah.

Shuqiang, W., 2014. Application research on developer-driven mode in BIM project: Taking a subway BIM project as an example [J]. *Constr. Econ.* 8, 10.

Smith, D., 2007. An introduction to building information modeling (BIM). *J. Build. Inf. Model.* 2007, 12–14.

Szabó, P., 2005. The Visegrád World Heritage Project: Problems and Tasks. *EPOCH and Individual Authors*, 91–96.

Tang, P., Huber, D., Akinci, B., Lipman, R., Lytle, A., 2010. Automatic reconstruction of as-built building information models from laser-scanned point clouds: A review of related techniques. *Autom. Constr.* 19, 829–843.

Telmesani, A., Sarouji, F., Adas, A., 2009. *Old Jeddah a traditional Arab Muslim city in Saudi Arabia*, 1st ed. King Fahad National Library, Jeddah.

Thomson, C.P.H., 2016. *From point cloud to building information model: Capturing and processing survey data towards automation for high quality 3D models to aid a BIM process*. UCL (University College London).

Thomson, C.P.H., Boehm, J., 2015. Automatic geometry generation from point clouds for BIM. *Remote Sens.* 7, 11753–11775.

Vergauwen, M., Van Gool, L., 2006. Web-based 3D reconstruction service. *Mach. Vis. Appl.* 17, 411–426.

Volk, R., Stengel, J., Schultmann, F., 2014. Building Information Modeling (BIM) for existing buildings: Literature review and future needs. *Autom. Constr.* 38, 109–127.

Waugh, L.M., Rausch, B., Engram, T., Aziz, F., 2012. Inuvik super school VR documentation: Mid-project status. In *Cold regions engineering 2012: Sustainable infrastructure development in a changing cold environment*. ASCE Publications, pp. 221–230.

Wong, K., Fan, Q., 2013. Building Information Modelling (BIM) for sustainable building design. *Facilities* 31, 138–157.

Wotton, H., 1968. *The elements of architecture: A facsimile reprint of the first edition (London, 1624)*. Folger Shakespeare Library.

Wua, T. C., Linb, Y. C., Hsuc, M. F., Zhenga, N. W., & Chen, W. L. 2013. Improving Traditional Building Repair Construction Quality Using Historic Building Information Modeling Concept. ISPRS-International Archives of the Photogrammetry, Remote Sensing and Spatial Information Sciences, 1(2), 691–694.

Xiong, X., Adan, A., Akinci, B., Huber, D., 2013. Automatic creation of semantically rich 3D building models from laser scanner data. *Autom. Constr.* 31, 325–337.

Yalcinkaya, M., Singh, V., 2014. Building Information Modeling (BIM) for facilities management: Literature review and future needs. In *Product lifecycle management for a global market*. Springer, pp. 1–10.

Yi, L., Feimin, S., Zhenyu, Z., 2016. Application of BIM in the full cycle metro project. *Appl. Mech. Mater.* 851.

Zheng, W., 2013. *A comprehensive analysis of building information modeling/model (BIM) policies in other countries and its adoption strategies in Ontario*. McMaster University.

Zhou, Y., Ding, L.Y., Luo, H.B., Chen, L.J., 2010. Research and application on 6D integrated system in metro construction based on BIM, in: Applied Mechanics and Materials. Trans Tech Publ, pp. 241–245.

Zlatanova, S. 1999. VRML for 3D GIS. Proceedings of the 15th Spring Conference on Computer Graphics,TC, Enschede, The Netherland, 28, 74–82.

Index

Note: Page numbers in *italic* indicate a figure and page numbers in **bold** indicate a table on the corresponding page.